# Zippy The TV Chimp

by

## Carole Womack

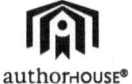

authorHOUSE®

*AuthorHouse*™
*1663 Liberty Drive, Suite 200*
*Bloomington, IN 47403*
*www.authorhouse.com*
*Phone: 1-800-839-8640*

*First published by AuthorHouse 12/11/2007*

*ISBN: 978-1-4259-7295-0 (sc)*

*Printed in the United States of America*
*Bloomington, Indiana*

*This book is printed on acid-free paper.*

# Contents

# Credits

I offer all of my thanks and appreciation to my husband, Don Womack, who knew and loved "Chinners". (That Zippy, which was the last chimp I trained, visited us often with his handlers, Jack and Fran Rinsky). Don's insight, writing and editing abilities and his loving encouragement kept me going.

Holly Gorrell, who originally edited the book for me, exhibited great patience and talent. Ann Finkelstein created the fabulous art work for the cover. Terry Wilson has kept Zippy before the public for many years on his website and has spent hours helping and encouraging me from afar. Brad Tirey, representative for the publisher, kindly encouraged me for four years to finish this book. My brother John Barbour and our friend Aubry Venable both found scores of "Zippy" pictures. Thanks to them you have some great images to look at.

I sincerely hope you enjoy the pictures and stories and refer your friends and relatives to this book.

May God Bless You!

# Introduction to Zippy the TV Chimp

Originally, I was just going to write a series of fun stories that happened with Zippy. If you travel all over the U.S. and other parts of the world long enough, you will have surprises. When you travel with animals, the trip becomes even more of an adventure. Since I decided six years ago to write down the stories, more and more of the memories and emotions have returned. Most of them were good, but as with life, some experiences are not so pleasant. I continued to write, and I included both aspects and wrote the truth. After all, if you make an omelet, you have to break some eggs. Into every day a little rain must fall.

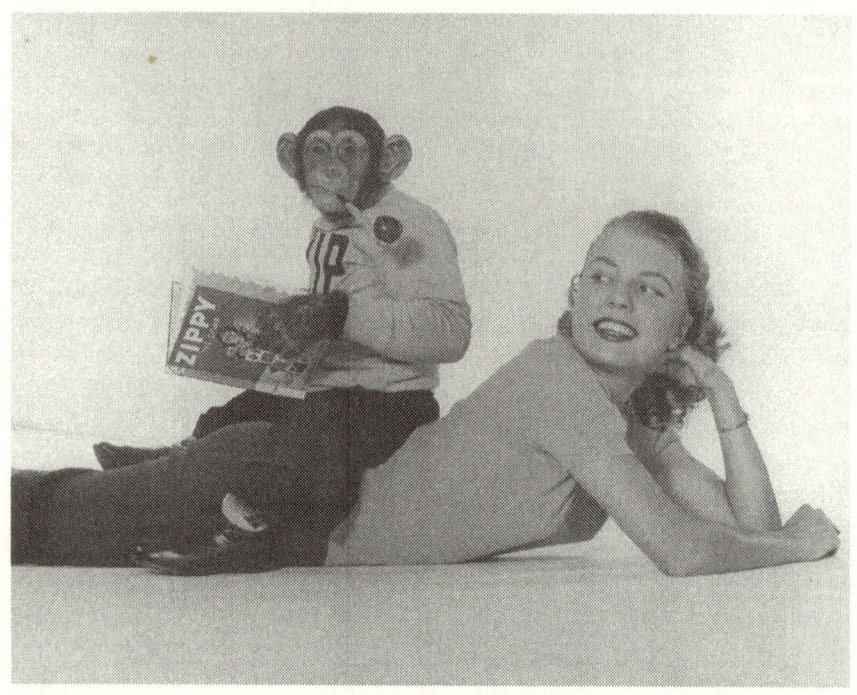

For twenty-three years, I had a "monkey" on my back

Although you will find a good number of pictures in the following pages, there must be thousands more that were shot by audience members, mall patrons, backstage guests or workers and people on the street. Zippy literally went everywhere and never refused a photo opportunity.

If you have some Zippy pictures or an interesting story, please share it with me. My next book about Zippy will pertain to *YOUR* stories. I have already been blessed by some beautiful childhood memories, shared by children who had Zippy dolls. These samplings indicate the far reaching effect a chimpanzee can have on a generation or two. I think a compilation of everyone else's stories may be just as fun, (and possibly even more interesting to me).

# Foreword
# By: Don Womack

In the 50's and 60's, animal acts were not as tightly controlled and monitored as now, and there were some incidents of animal attacks on humans. Some resulted from handler's carelessness; sudden movements by an audience member that were interpreted by the animal as aggressive; eye to eye contact perceived as challenging; appearance of fear indicating weakness; even hormonal conditions and monthly cycles. Knowledge and experience gained over the years has been available for trainers/handlers so that the potential for an attack can be recognized and an incident avoided. State and federal authorities seem to over-react in the extreme when there is an incident, but the resulting rulings usually establish guidelines which ultimately protect the public.

If you work around animals of *any* kind long enough, you *will* be bitten or attacked in some way. We must remember, after all, that loveable creature is an animal. Carole was never attacked, and it is attributed to understanding how chimps think, establishing authority immediately and never losing that authority by showing any weakness. She knew <u>never</u> to let a chimp's testing and challenging actions go unanswered. Control would be permanently lost if that happened. Carole was threatened once when she was sick and actually did lose control of that particular chimp. Carole's affection for her animals and yet her ability to maintain authority over them at the same time is phenomenal. I saw it with the chimps, and I see it with our Huskies and exotic parrots. Her rapport with animals is uncanny.

An adult chimpanzee is known to possess the strength of seven grown men. Armed with this knowledge, a 117-pound woman should not attempt to physically control a 70-pound chimp. Injury would be inevitable. She should know the risks, danger signs and subtle hints the animal will give and not put herself or others into an unhealthy situation. Animals do animal things. Their instincts are alive and well, regardless of their training. They are still animals. Chimps are very intelligent, but they are still chimpanzees.

Because of the potential for attack and to reduce injury in case of an attack, each new chimp's teeth were pulled. The removal of Zippy's teeth helped to reduce the risk to Carole, as well as anyone else. Carole has little scars on both arms from bites, most resulting from a "battle of wills" with baby chimps, (before extractions), and not from blind aggression on the chimp's part. Even without teeth, Zippy enjoyed eating apples, (and anything else he saw us eat), and could cause a significant bruise. An attack by Zippy would probably have been frightening, but without teeth, injury would be minimized.

The various Zippys successfully worked with hundreds of personalities on and off the screen and in all types of surroundings. "Chinners", one of the best Zippys ever, accompanied by his handler, (Jack Rinsky), appeared on the Johnny Carson Show for the last time in the late eighties. Carole and I watched the TV screen as Zippy sat in his chair and amazed the audience with his ability to tie his shoes, button and unbutton his shirt and use sign language. (Chimps' "thumbs" are not like ours and are said to be an identifying factor in non-human primates and a limiting factor for the dexterity of chimps, but Zippy didn't know that.) Jack remained courteous and focused and kept Zippy busy with his feats. However, these high achievements were overshadowed by Carson's continued exclamations, "That chimp bit me!" Nothing happened on camera and Carson still had all his fingers, a nose, both eyes and ears, there was no blood and his hair was still impeccably quaffed. Apparently *something* happened backstage to either scare Carson or inspire him to make such a claim. (Remember, if there was a confrontation in the Green Room, Carson was the only one with teeth.) His fear of animals brought him a lot of laughs throughout his career, but this time Carson rode the theme into the ground ...along with Zippy's future. Zippy was convicted in the "court of public opinion" by, "That chimp bit me," although the evidence was overwhelmingly to the contrary. I turned to Carole and said, "Well, that did it for Zippy".

Carole was Lee's third wife. She was not around at the earliest introductions of Zippy or for the demise of Zippy in the eighties. Carole trained all but the first two of the Zippys. (One was "trained" by Lee and the other by Ralph Quinlan.) Carole, as a result of her rapport with the chimps, was able to build Zippy into one of the most visible and publicized attractions of the time. You have to be doing something right when all the leading talk show hosts and agents want "Zippy". Zippy was everywhere and even dined at some of the country's best known restaurants, with his napkin tucked into his shirt and excellent table manners. Everyone was impressed. Of course, he would not be allowed into a food service establishment now. When you live with animals, crazy things happen.

# Biography

**"The talents my Lord gave me are His gifts to me.
What I do with them is my gift to Him."**

Carole was born in Seattle and she had show business in her
blood. Carole's father, John Barbour, was a concert violinist,
first chair and assistant conductor of the Seattle Symphony Orchestra. He met Carole's mother, Mary Louise, when he was
looking for an accompanist to tour with him. She was a very
talented concert pianist. She got the job and married John.

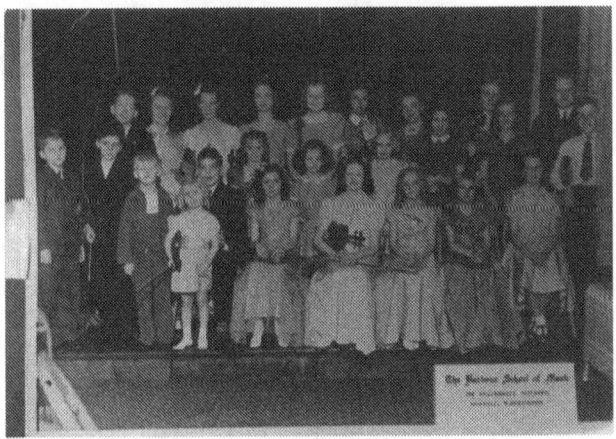

That's me the little girl up front with the violin

Carole got into show business at an early age. John built violins as a hobby and he created an eighth-sized violin for Carole. He taught her to play that little violin when she was three years old. Mary Louise made evening dresses for Carole, which matched her own gowns. Carole played her little violin on stage between her father's solos. Carole later took piano, dancing, and acting lessons, but discovered that her greatest talent was singing. She was trained to be a Broadway performer.

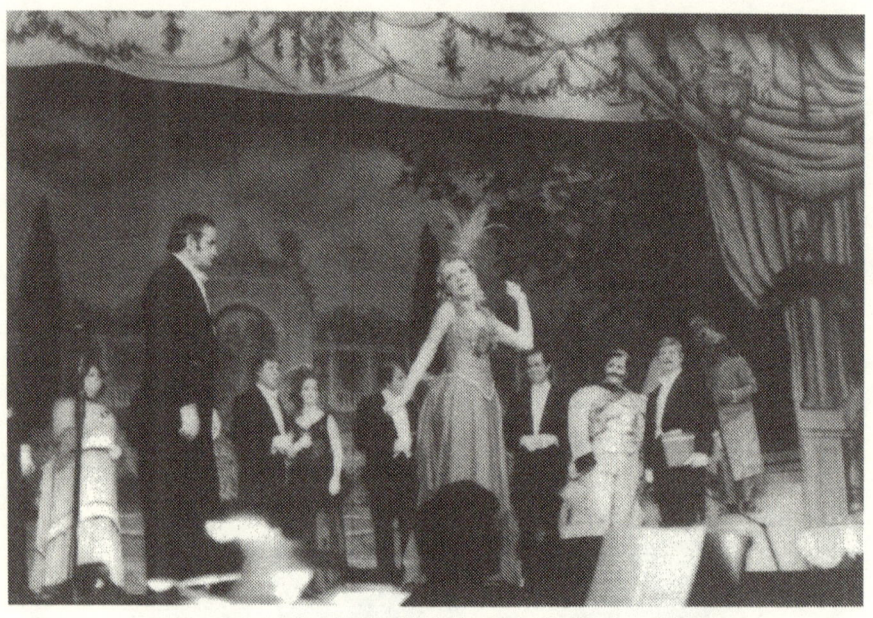

Carole (Adele) sings "The Laughing Song" from
Die-Fledermous

Carole spent her mid-teen years in Germany studying opera, languages and drama. Her opera coach, Hildegard, was the diva of the Nurnberg Opera. Carole spent dozens of nights in the front row center seats watching opera and learning. Eventually, she auditioned for Heir Houser, who was then the Impresario of German Opera in Hamburg. He was the counterpart of Rudolph Bing, the Impresario for the Metropolitan Opera. After auditioning with two arias, Heir Houser said, (in German), "How does all that voice come out of such a little toothpick? At the time she was about 117 pounds and 5'3", but she always had a powerful voice, even with her diminutive size and youth. Heir Houser offered Carole the opportunity to sing with the Hamburg Operetta Company, which he had just created. Unfortunately, it was not meant to be. Carole's family had to return to the USA, and at seventeen, they felt she was too young to be left behind in Germany alone

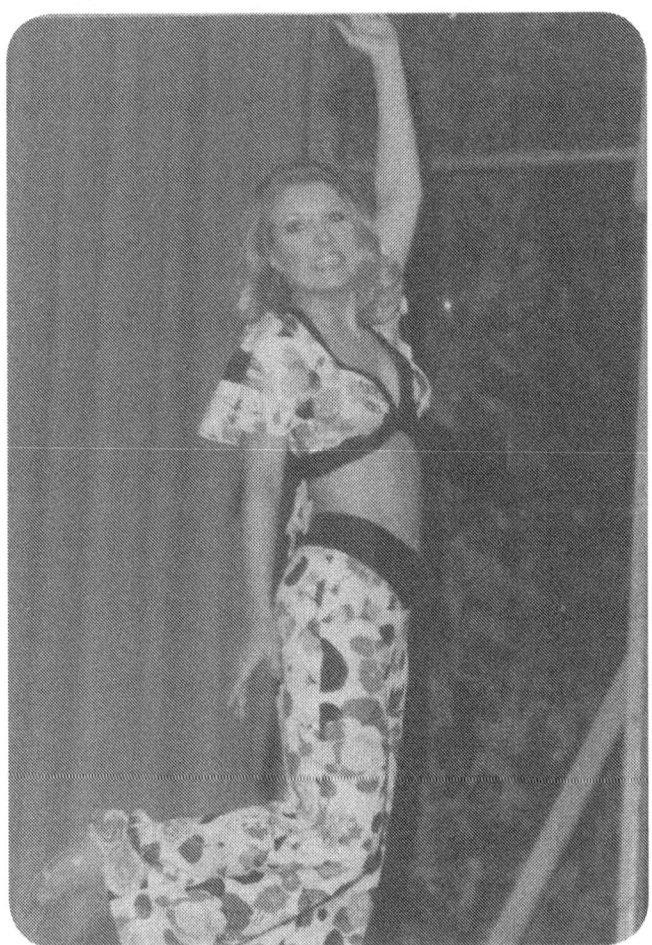

Carole sings as Lois in Kiss me Kate

Carole continued her study of opera back in New York, but her coaching lessons were very expensive. So she auditioned on Broadway hoping to land a job that would pay for her studies. A director/producer from Paramount Pictures took Carole in hand and began grooming her to be another Jane Powell or Kathryn Grayson, etc. One day Carole auditioned for the road show rendition of "Sound of Music" with Florence Henderson. Her audition number was "Porgy" from "Porgy and Bess". Carole felt confident that she had "nailed" that song, and it was confirmed. Believe it or not, Florence Henderson walked out on the stage, put her arm around Carole, and said to the group of men in the middle of the theatre, "I want her with me!" A gentleman stood up in the dark seating area and said, "Send her to the attorneys to sign a contract!" The next girl in line asked if she could audition using "Porgy" also. Florence turned to her and said, "No, no one can sing it like she did." Wow! What a compliment! Carole considers that day as the turning point in her life. She was never so thrilled, and even today is very thankful to Florence for the part she played in launching Carole's career.

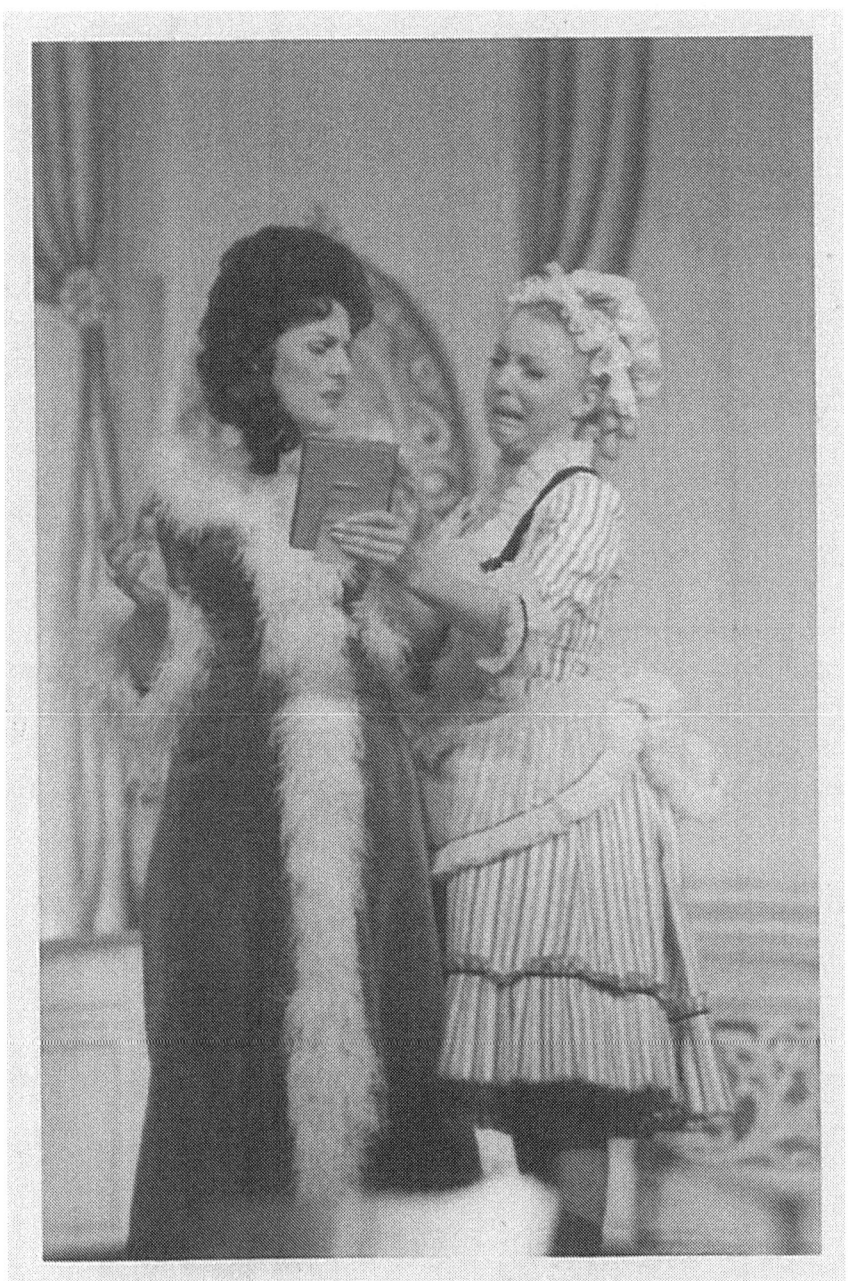

Carole sings "The Letter Song" as the Maid in Die–Fledermous

Long before that last audition, Carole's mother had discovered an ad in the New York Times that said, "Girl wanted to work in show business with animals. No experience necessary." Mary Louise saw this as a great opportunity for Carole to meet agents and show people who would help to advance her career. Coincidentally, Carole's father was raised on Long Island, and he took her to the audition to work with Zippy. Carole got the job, and three days later she performed her first personal appearance with Zippy at Hecht's Department Store in Washington, D.C. Carole's first TV show was unfortunately the very last Howdy Doody show to include a performance by Zippy. Not long after that show Carole married Zippy's "father", Lee, (Carole was Lee's third wife) and was at another monumental turning point in her career. She was faced with making a choice between her singing career, and working with these fabulous little apes. Carole's future was to become an animal trainer and variety show performer. However, she did not completely leave her singing career behind. She performed on Long Island in several big productions with full orchestras, huge stages and audiences. She was Adele (the second lead) in the operetta "The Fledermous"; the lead, Julie, in "Carousel"; and Lois/Bianca (the second lead) in "Kiss Me Kate". She currently sings solos in her church from time to time.

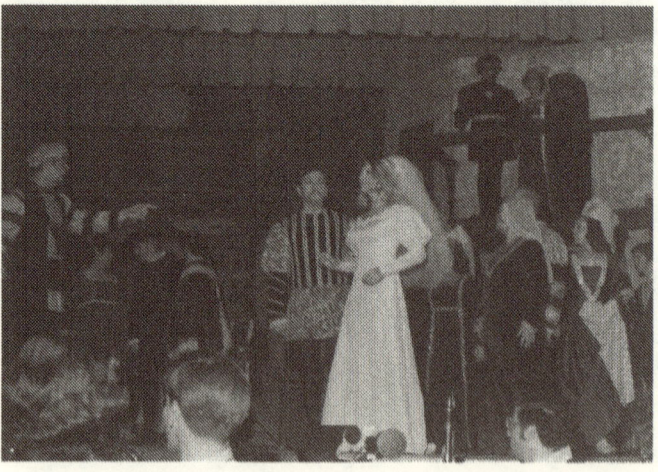

Carole (Bianca) in the "wedding scene" of Kiss me Kate

After training the first real Zippy, Lee never trained another chimp. Carole trained all the subsequent chimps, as well as the couples who were sub-contracted to work with them. The truth is that most of the chimps did not like Lee and would not obey him very well. While on the stage, Lee would talk and handle the microphone, and Carole would handle the chimps. Lee had many different interests. After a few years, Lee lost interest in Zippy.

Carole went on to train and perform with exotic parrots, during which time she met and married Don R. Womack, her present husband. Don says that Carole suffers from tunnel vision. When she sets her mind to do something; nothing gets her off that track. Perhaps that is what makes her such a great animal trainer. She tries to train beyond what the animal can do; and they do it, sometimes to her own amazement. This is especially exemplified in the Speidel Watch Band Commercial chapter and later "The Ferris Wheel", which involves the parrots. Together, Carole and Don developed "Alfie Cockatoo and his Fun-Lovin' Macaws" and raised Don's three daughters from his previous marriage. They have spent 26 years together working as a team to date.

Carole is planning a third book which will be about "Alfie Cockatoo and the Fun Lovin' Macaws". This book will tell the story of a fabulous "variety act" that traveled and entertained audiences for twenty-six years. The birds were trained by and performed with Carole and Don in over 40 states in the USA, Canada, Central and South America. It will be an exciting read with plenty of pictures.

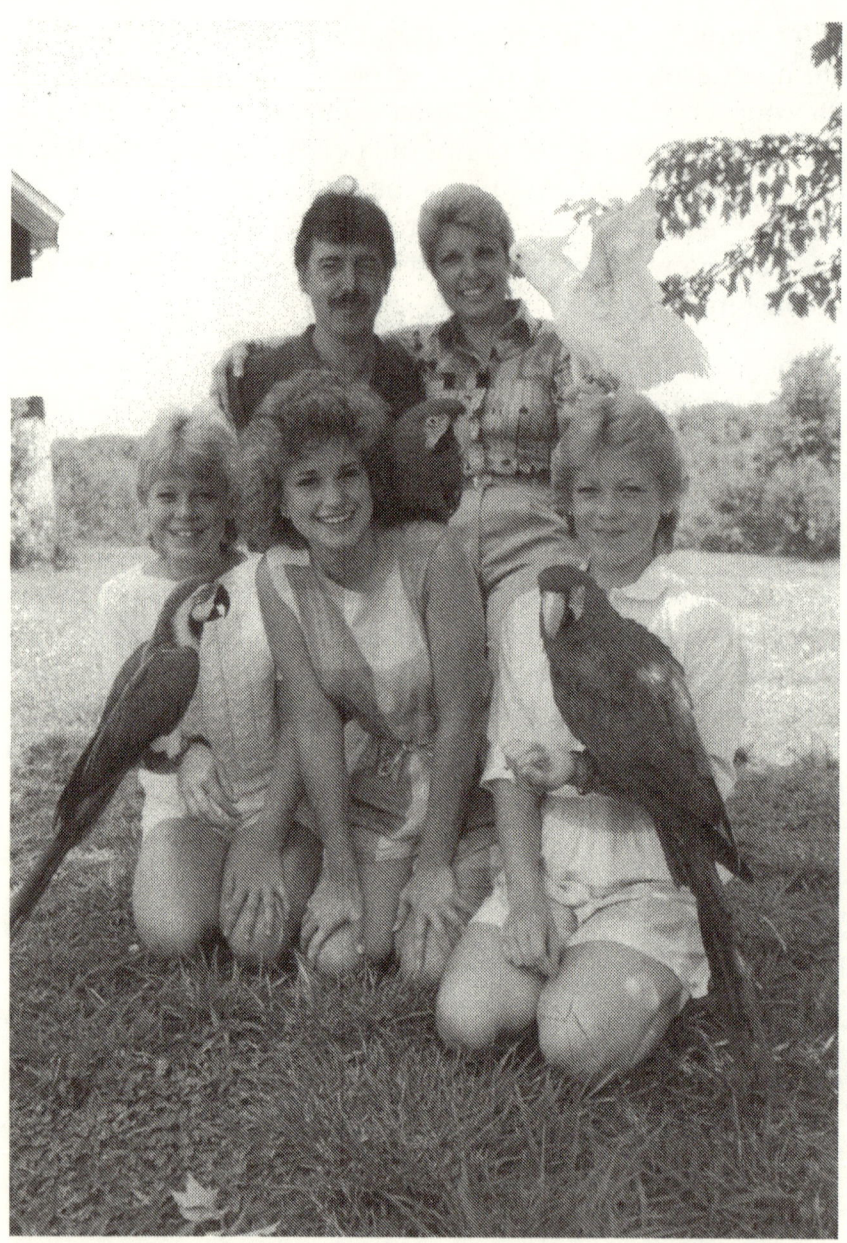

Our family with Alfie Cockatoo and His Fun-lovin' Macaws

# How I Met Zippy

My dad had brought home the New York Times, and my mom was intrigued and started looking through the want ads for me. As it happened, I was looking for another job at the time. There it was, "Wanted, attractive young lady to work with animal act. No experience necessary." I was less than thrilled at the prospect. "Mom, I don't want to work with animals. My entire life, I've studied to be a Broadway star!" Her response was quick, "But don't you see? You'll be working around show people, get a chance to meet agents, and then you'll move right up. Give it a try, call and see what it's all about.

If nothing else, you have to admit the ad is curious. It will only take a moment." So I called, and when I learned that the job was to work with a chimpanzee, I was fascinated. Then, when they mentioned the location, my dad got interested, "That's where I went to high school. I could take you there and perhaps meet some of my old school friends." The time and date for the audition were set.

I wore my prettiest dress for the trek, which was several mornings later. My dad and I drove up to and parked in front of the house. House? It looked more like a mansion to me! I later found out that it had 5200 square feet of living space. English Tudor style homes have an elegance all their own, and with English ivy gardens, stone retaining walls and huge trees, I felt intimidated as I approached the very large mahogany door. With a quick glance I noticed two identical Cadillac's in the driveway. After a moment's hesitation, I rang the bell, and nervously waited.

The door opened and there stood this adorable chimpanzee all dressed up like a little boy!

He took me by the hand and led me through the mahogany and stone entrance hall into a huge living room. There in the middle of the room was an "L"-shaped white leather couch; 18-feet long with a matching chair. Seated were two men and a woman with a note pad. We said hello to each other, but that was about it. The chimp started pulling me around the couch, indicating that he wanted me to chase him, so I did. Round and round we went. The more times I changed directions, the louder he would laugh that deep throaty laugh a chimp does when he is genuinely having fun.

I was only 17 years old at the time. A lifetime of dancing and singing lessons provided a strong, solid body and I wasn't easily winded, but the chimp eventually outlasted me. Finally, I fell into the chair and the chimp dove into my lap. He placed my hands under his jaw and someone said, "Zippy wants you to tickle him." I did, and wow, did he eat that up! I think every spot on his body was ticklish and soon he was ready for a rest, on my lap of course. I suddenly realized that I had hardly spoken to anyone, so I asked in a rather timid voice, "Would you like to know about me?" The two gentlemen were very pleasant as they asked the general questions and the lady wrote down my answers.

They kept it pretty brief and soon I was back in the car where my dad had been waiting. When he asked me how it went, I said, "I think I made a fool of myself. All I did was play with the chimp the entire time. I also think I gave them the wrong telephone number." He told me not to worry about it, and said, "You can call them from the restaurant and make sure they know how to reach you." I asked, "But won't that make me look stupid?" He smiled and chided, "Well, you didn't want the job anyway, so why worry?" I retorted, "I don't know, but I think Mom was right. It could be a good way to make important contacts for getting into show business, and frankly, it looks like fun. The chimp is fabulous, and he really seemed to like me." I pondered all through what little lunch I ate and finally got up enough nerve to go to the phone and call to see if I had left the correct number. When I called the gentleman said, "We're glad to hear from you, Carole. We would like you to have the job. Out of the 200 girls that were interviewed, you were the only one that had a rapport with the chimp. How soon can you start?"

It turned out they were right about the rapport. For several days, I got to know the chimps. There were several of them and each one seemed to like me. At the time, I had no way of knowing that I had a knack for working with animals, that special authority that causes them to respect and love you. Almost from the start, I found new things to teach them. The challenge was delicious and I adored teaching those little guys. Later, I would go to toy stores and FAO Swartz to find new toys and props they could use. A seven-minute act soon stretched to 25 minutes, with behaviors to spare. (Note: We did wild and remarkable things with the chimps, and they did some really funny and fabulous things on their own, which will follow in the chapters to come. As a born-again Christian, I'm not comfortable with bragging. However, the Lord gave me a great talent for training animals, and I would like to share some of my experiences with you).

Now, let's go back to the beginning. Zippy's daddy was very aware of my lack of appropriate clothing. Being the daughter of an Army Sergeant and one of seven children, I was not blessed with many new clothes and I was about to be traveling around in very nice circles. (I just didn't know that yet). So, Zippy's daddy took me shopping. We went to some nice dress shops and boutiques in town and found one that had dresses that I loved. I tried on one after another and everything just seemed right. I modeled them for Zip's dad and asked him, "Which one shall we get?" He replied, "All of them, of course. Now let's go get you some shoes."

Only three days later, we were on our way to Washington, DC, to perform in a Hecht's Department Store. Since I had never seen the show before, I asked what I should do. I was told to hand over whatever prop was asked for, as it was needed. It seemed that I was to be a prop girl. Well, I had been performing on stages since I was 3 years old, so it only took two shows for me to work my way right into the act. I began "*MY*" new act by singing a few songs, to and with the kids, and then I gave Zippy a flamboyant introduction. (Whenever we could, we had Zippy make an unusual entrance. If there was a curtain, he would peek under it, stick his foot out, or come skating down an isle as fast as he could and leap up onto the stage.)

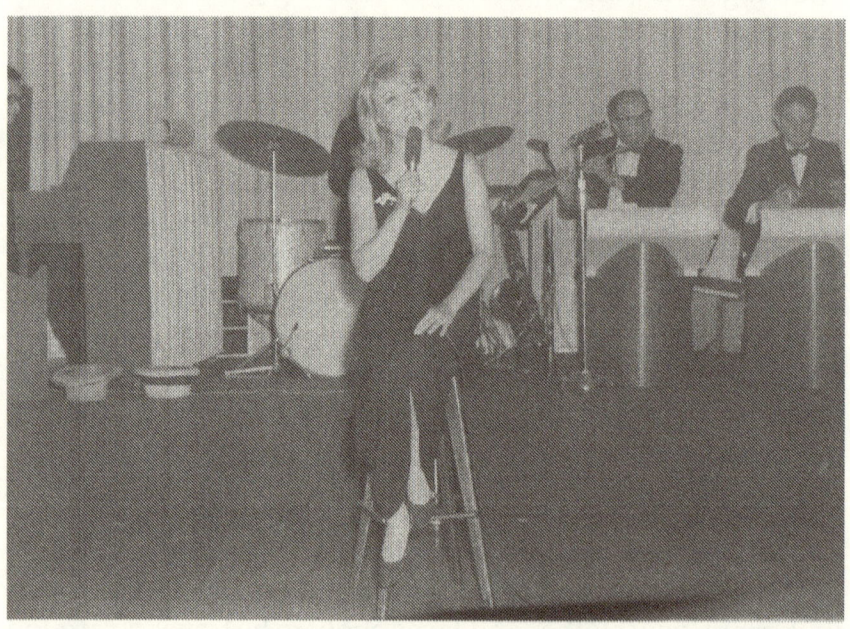

Carole entertains in New York

# Beginnings

It all started when a psychology professor at Loyola University in New Orleans, happened to mention in one of his classes that he owned a chimpanzee. One of the students present that day, Lee, became fascinated by the professors' stories and set out to find a chimp of his own.

Searching for a chimpanzee, while knowing absolutely nothing about them may not have been the wisest of actions, but it eventually led Lee to Florida for a "bargain" in a very big, black-faced chimp. Lee placed a leash around Big Joe's neck, put him in the passenger seat and headed home. As luck would have it, Lee was stopped on his way back home in one of the infamous Georgia speed traps of that era and was taken to the judge's quarters, (behind the local gas station), to be fined for "speeding." Since he couldn't just leave Big Joe alone in the car, Lee brought him in and had him sit in the large stuffed chair that was in the corner of the room, where the chimp could be comfortable while they waited. As soon as the fine was paid, in cash of course, Lee escorted Big Joe back to the car. However, the big guy had left a payment of his own in that stuffed chair; in much the same way as the orangutan "Clyde" of "Any Which Way But Loose" fame would show his appreciation for the uniform many years later.

Carole Womack

I do not know the fate of Big Joe, only that a smaller, white-faced chimp was purchased at a later date, and eventually the third chimp, which proved to be the charm, the original Zippy. I'll never know why Lee went to the French Quarter for work. After trying several strip clubs he finally landed a job at the Gunga Den, where the show consisted of 9 strippers plus Zippy. They performed a number of these shows each night. One particular evening, Buffalo Bob from the Howdy Doody TV Show happened to be there. What was Bob Smith doing in a strip joint? For that matter, what was Lee doing in a strip club? Lee was studying to be a priest! We may never know the answers to many of these questions, but it all worked out well for Zippy and Lee. Zippy stole the cherries out of Bob's drink, and Lee headed to New York with a 5-year contract for the Howdy Doody Show.

After arriving in New York, Lee and Bonnie, (a stripper he had recently married), found a camper-trailer park in nearby New Jersey and pulled their newly purchased trailer in beside an established skating act, Ralph and Kathleen Quinlan. The two couples soon became fast friends. The trailer had no heat and New Jersey gets pretty cold, so Ralph helped Lee install a heating system.

While Lee and Bonnie worked on the Howdy Doody TV Show and traveled with the road show, Ralph and Kathleen cared for the baby chimp, Lucky, whom Lee had recently purchased. Ralph and Kathleen fell in love with the little chimp, worked with him, and soon Lucky became a second "Zippy". Being professional trick skaters themselves, Ralph and Kathleen soon equipped the chimps with professional skates of their own. That is how the chimps became such fabulous skaters.

A few years later, Bonnie left Lee, at which time he applied for a girl to work with him. That was when I arrived on the scene. While Lee and I were handling the television shows, dealing with agents, contracts and making local appearances, Ralph and Kathleen handled the roadwork.

They traveled all over the country making personal appearances. They drove an identical Caddy to Lee's, and most people believed the chimp with them was the original. It's sort of like Lassie. (There were several Lassies used on television and for personal appearances). Several years later, Kathleen left us due to health reasons. She and Ralph eventually divorced, and Ralph went on to train dolphins and in general became quite an entrepreneur.

Although live television shows were the specialty, the number of live shows was diminishing and taped shows increasing. Regardless of the format, Zippy could turn a mediocre show into something spectacular.

While Zippy was contracted to the Howdy Doody Show, another contract came through for Zippy plus one other chimp to play the part of Cheetah on the upcoming Tarzan Movie.

Two of our chimps, Zippy, (handled by Lee), and Lucky, (handled by Ralph), were in a Tarzan movie called, "Tarzan's Hidden Jungle," starring Gordon Scott and Vera Miles. These two fell in love during the filming of the movie and were married later, (the people, not the chimps). The movie was not a big success. Many reviews said the plot was overly simplistic. However, Lee and Ralph were highly complimented by the film's producers, because as the chimps' trainers, they had saved the film's producers a tremendous amount of money on film and time. The chimps had performed as expected the very first time, every time.

At this point, I will share two funny stories that resulted from this film. Zippy's "daddy," (Lee), happened to be an excellent photographer. He was shooting his own private movies of the sets and parts of scenes for later amusement. While filming an episode at a lake where Jane, (Vera Miles) was supposedly bathing nude, her bathing suit kept showing and this constantly required more "takes." Finally, Vera Miles decided she had had enough of it all, stood up, and pulled her top down saying,

"This will end that problem!" Lee swung his camera around to film this priceless shot, but he had run out of film.

During some publicity photos, Gordon Scott was asked to hang from a beam and have Cheetah, (Zippy) hold on around his neck. While Tarzan was hanging from the beam as asked, Zippy's arms were placed around Gordon Scott's neck. Zippy began to lose his grip, and as he slipped down Tarzan's body. He took the loincloth with him, and Gordon Scott was left hanging au naturale.

# The Disappearing Chimp

The original Zippy was quite a phenomenal chimp. It was as though he could sense what he was supposed to do on a television show. At that time, most of the shows were filmed live, so it had to be right the first time. There was no chance for a re-take or do-over. The first Zippy also had a unique and curious side to him. He would disappear. Fortunately, it never happened on live television, but at other unexpected times, he would pull his disappearing act. Why don't I start from the beginning?

I was into my job about three days when we took off in the gorgeous, solid white Cadillac for Washington, D.C., to work at a Hecht's Department Store. We checked in at a hotel and they gave us the top floor (the loft). It was huge and it had several beds, a kitchenette, and dormer windows on all sides. Upon discovery of the fully-equipped kitchenette, we headed out to the grocery to buy supplies and a cookbook. I loved to cook, but remember- I was just seventeen years old and had no idea how to prepare some of the food we were buying.

As I was diligently learning how to steam artichokes, Zippy's dad relaxed with a book and Zippy played with his toys on the bed. All was going well as I was thinking, "This is the life."

All was new to me, and life on the road seemed so very glamorous and fun. As I glanced over to the couch, I noticed a book had fallen onto the floor and snoring was coming from that couch. Zippy's toys were neatly on the bed, but Zippy wasn't there! Oh dear! What had he gotten into now? Quickly, I woke up the snoozer and together we hurriedly searched for Zippy.

The door was bolted, so he couldn't have gone out. Every nook and cranny was as it should be.

Where in the world could he be? We started from scratch and looked everywhere again. We were totally at a loss until I noticed an open dormer window. I ran over to the window, pushed it all the way open and looked around. There was Zippy, sitting on the roof, quietly and calmly watching the traffic below, *way* below I might add. It took a little coaxing to get him back in, but all was soon peaceful again in our loft overlooking Washington DC.

The next day at the department store, I was being shown what to do during the show, when suddenly we missed Zippy again. It was easy for us to follow the direction of his escape. Not too many people can miss a lone chimpanzee, dressed as a little boy, going down an escalator. We quickly followed the pointing fingers of the shoppers until we came to the china department.

Our hearts skipped a beat. A chimp in a china shop is a genuine cause for concern. Just imagine the damage that could be done! Our worst fears were not confirmed. The sales lady had Zippy sitting on the floor in a corner, while showing him some pretty things. Both seemed to be having a great time. Zippy was fascinated as he rang a china bell, trying to figure out from where the sound was coming. He was entranced with his new-found friend. We took him by the hand and gently led him upstairs amidst all the cheering patrons. The next three days we had huge audiences. Many had heard the story and wanted to see the chimp from the China Department. Evidently, Hecht's

Department Store was pleased with the draw and publicity. We entertained at many of their stores thereafter.

One of my favorite stories happened during a tour of theaters in upper New York. This chain of theaters was featuring a special for kids, a day at the cinema, so to speak. It included free candy, many cartoons and a live performance on stage, featuring the one and only "Zippy the Chimp". I believe each child was also given a picture of Zippy. We would usually arrive while the cartoons were showing. We had to set up our props in the dark behind the screen and had to stay rather quiet while we waited. During a conversation with the theatre manager, we discovered Zippy was missing. The stage door to the outside was open, so that was the natural place to start searching. What if Zippy got out on the highway? That was something I didn't even want to think about! As the men frantically tore outside, I took another route. Knowing how much Zippy enjoyed watching television, I walked up and down the aisles looking for an ugly kid. Nothing on the first level caught my eye, so I went up to the balcony. As I started down the first aisle, I spotted him. There was Zippy: sitting in the first-row aisle seat, next to another child, watching the cartoon just like the rest of the children. Not one of the kids had taken any notice of him. I took him by the hand and walked him out with no one the wiser.

Pretty much the same thing happened in a home for older gentlemen. We were talking over details of our performance with the manager of the establishment, when suddenly Zippy took off on his skates. (Now you must realize that Zippy was no slouch at skating. His skates were professional and he could skate as well as any professional in the business. Zippy was totally at home on his skates. He wore them everywhere we went. It was a common sight to see Zippy, dressed as a boy, skating between us as we walked up 5th Avenue or Broadway, or Main Street in Any Town, USA. In fact, while we were in Washington, DC, we visited several sites. We actually took Zippy in to view congress while it was in session. The Lincoln

Memorial was Zippy's favorite. Perhaps you can recall the myriads of steps one has to climb to visit Honest Abe. Well, Zippy had a wonderful time skating up and down those steps as fast as he could go.

Getting back to our story, Zippy tore through the men's home on his skates while I ran fast as I could after him. At one point I lost sight of him, but when I saw a sign over a door to one of the rooms, I knew where to look. The sign read, "Television Viewing Room" and the room was set up like a theater, with chairs and comfy couches arranged in rows. Zippy had found an aisle seat near the back and was watching some TV. I took his hand and Zippy quietly left with me. Not one of the gentlemen ever saw him.

Basically, Zippy got bored whenever we had to stand and talk. He knew that if he tried, he could find something elsewhere to entertain himself and he went searching for it. He was strong-willed and maybe had an attention deficit, but never was he malicious. Although his escapades were scary at the time they happened, they became something to laugh about when we knew Zippy was safe once again.

Often we would do publicity promotions before an engagement to attract people to the shows. Usually it would be a local television or radio show or a visit to a newspaper office. One day, somewhere in the USA, we were waiting to go on a news show. A quick rehearsal showed Zippy where he was to work and with whom. Most shows were impromptu and it was up to the anchorman to work with Zippy to create something fun. It always worked out well.

As we were waiting to go on, can you guess who disappeared again? You got it! Where in a studio could a chimp go? It wasn't that big! I pushed against the heavy floor-to-ceiling velvet curtain and realized there was no solid wall behind it. I found an opening and discovered a cooking show going on live on the other side of the curtain.

Now picture this: The star of the show is appearing on <u>live</u> television, when suddenly a chimpanzee, dressed as a child, jumps up on a stool next to her working counter, picks up a fork and starts sampling the food she is cooking. Some people might have freaked out, but this gutsy woman calmly said, "Hello there." Noticing the "ZIP" on his shirt, she said, "And I guess your name is Zip", and played right to the chimp. What a marvelous job she did! We stood back and let them go. She kept on cooking, but directed her conversation basically to Zippy, showing him how to stir and treating him just as though she were teaching a child how to cook. Zippy used the utensils she handed him like a pro. It was one of the cutest shows ever. Unfortunately, it was live, so there was never a record of it. I can't remember if we ever did the news show or not. Whether we did or not, it would have been overshadowed by the cooking show.

# Fun and Famous Television Appearances

Zippy never took off when performing a live performance or on a television show *except* for one delightful time on "The Phil Silver's Show." He didn't leave the stage or the building, but he left his assigned chair. It may sound like a loss of control, but Zippy only moved from one desk to another. His ad-lib seemed totally appropriate and scripted and was the highlight of what Phil Silvers and all the directors and producers will agree was the funniest show Phil ever did. The episode was called, "The Court Martial of Harry Speakup."

The story progressed around the theme, "Indoctrinating men into the army at super speed." Through comedy glibs, Zippy finds himself in line with the men being indoctrinated and is overlooked because of the rush. As the man at the desk calls out the names of the men passing by, another soldier is writing them down. The next man in line is supposed to be Harry *somebody*, but it is actually Zippy and naturally, Zippy doesn't answer when a name is called. The man yells, "Harry, Harry, speak up!" The other soldier writes the name down as "**Harry Speakup**" and thus the name is established. Zippy is pushed through the measurements of height and weight. Then he opens his mouth and the dentist calls out, "over developed incisors!" Several more such fast-paced scenes occurred! The recruits are taking a written test as they sit at individual desks. Zippy's paper is switched with that of another recruit and Zippy passes the test!

The scene with the foot doctor is hysterical! The doctor is examining feet (and only feet), one after another. When Zippy's hairy feet appear before him, the doctor takes off his glasses, wipes them and says, "You view feet all day long, this is bound to happen".

Everything happens at a fast and frantic pace, and suddenly it is discovered by the officers that Harry Speakup is indeed *not human*. Now what do they do? He passed all the tests with flying colors. He has been successfully inducted into the Army! Now, how do they get him out? They decide upon having a court martial and Phil Silvers becomes the lawyer defending Zippy.

Please understand that this whole show is done in front of a full, live audience. It is a 30-minute show interrupted by commercials and everything is done as though it were a stage show. Men were scurrying everywhere and, as usual, there was very little rehearsal for Zippy. Yet, Zippy was magnificent!

The last portion of the show was the actual court martial. Zippy mostly sat at a desk by himself. He was drawing with a pencil and paper and trying to keep his army hat on while the other actors played out the comedy courtroom scene. Suddenly, Zippy simply became tired of sitting at the desk. He pushed his big chair back, swung out of his seat and headed for a prop telephone positioned on a table toward the back of the courtroom. Then Zippy picked up the phone's headset, put it to his ear and started spinning its rotary dial.

Phil Silvers played right along with Zippy's improvisation. He turned his body, slowly following Zippy's direction; and when Zippy picked up the phone, Phil struck his own hand to his head and said, "Oh, no! He's calling another lawyer!" It had to be one of the most fabulous ad-libs in television history.

Absolutely everyone broke up laughing hysterically with the exception of Phil Silvers (absolute professional that he is), who managed to continue with a straight face. Zippy hung up the phone, went back to his place at the desk and the court martial continued to a happy conclusion. The Court Martial of Harry Speakup made television history and became an instant classic! At the time of this writing, you can use your internet to see the last portion of this show. Simply search for "Harryspeakup" and then watch the video.

There is also a clip of Zippy on the Ed Sullivan Show under "Ed's Amazing Animals". This clip shows Zippy jumping (while wearing his famous skates) through a Hula Hoop I am holding. He jumps through it beautifully and glides to the nearest TV camera and sits on it. Then he skates back to Lee for a strong send off and this time I hold the hoop higher. Zippy repeats this once more as I really hold it high (you won't believe it!). At this point, Zippy feels he has done enough, so he returns to the high chair and begins to remove his skates. He throws one on the floor and as he begins to take off the second skate, Ed Sullivan walks up with a lit cigarette. Zippy drops the skate and it lands on Ed's toe. Now these skates are <u>very</u> heavy. Ed winces in pain and hands Zip the cigarette. Zippy takes a puff, throws it on the floor and stomps it out. It is an adorable bit of history.

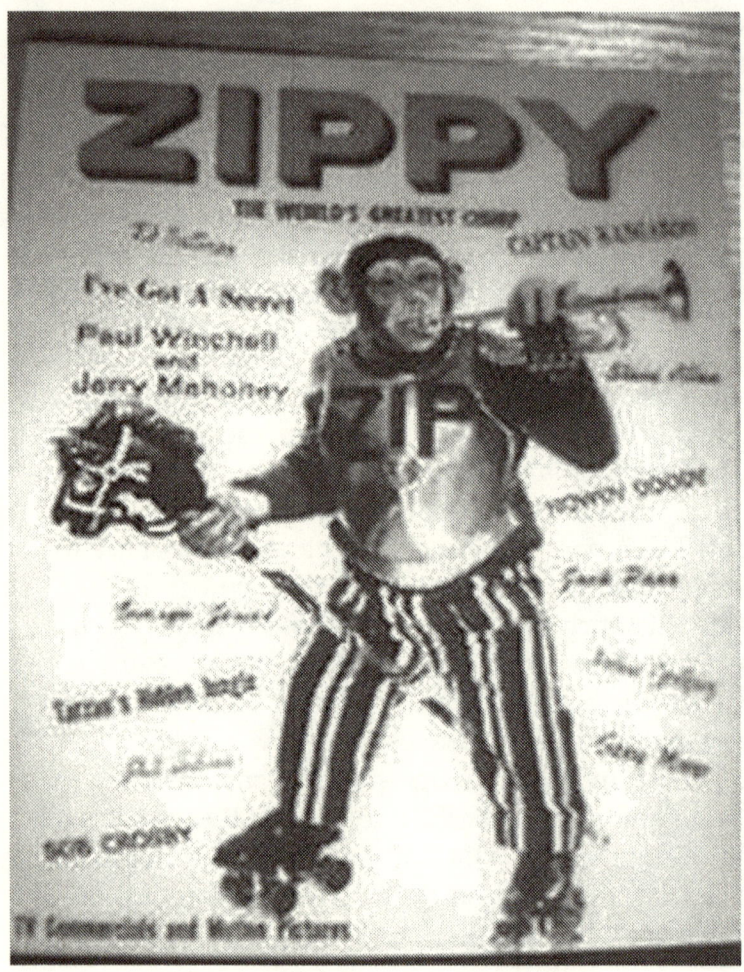

Zippy's first brochure

Captain Kangaroo was always a favorite show and Zippy appeared on it regularly. As you may recall, the show usually started with Captain Kangaroo walking into the room and hanging his keys on a peg. When Zippy was on the show, he wore a costume fashioned after the Captain's and carried the Captain's keys over to the famous peg and hung them on it.

Mr. Greenjeans, the Captain and Zippy had a wonderful rapport between the three of them. This laid-back show picked up ratings every time Zippy appeared. The multi-talented Zippy rode his bike, skated, painted, etc, etc. Whatever the writers wanted him to do; he did and did it well.

"The Gary Moore Show" was another television show that used the chimps on a regular basis. Gary was grand with the chimps. This morning show is very much remembered by housewives across America. So many wonderful skits were arranged, it is hard to remember that any one was better than another. They were usually all good and Zippy's fame was at its height with "The Merry Mailman", "Ring-A-Ding School", "Super Circus", "The Price Is Right" and numerous others television shows.

I don't remember how many Ed Sullivan Shows we did, but I do remember when "Ham", the chimp astronaut, went up in a spaceship. The producers called us and wanted Zippy immediately on the show. They wanted Zippy to portray Ham, but Ham wore no clothes. We objected because Zippy was famous for wearing clothes and especially known for the ZIP across his chest. Well, we didn't get very far with our argument. All they wanted was a naked chimp to walk out of a fake space capsule to Ed Sullivan, who was to pick him up, receive applause and carry him to the wings, back to one of us. This seemed simple enough, so why not? Why not indeed! Zippy didn't like being naked and was very uncomfortable being naked in public. He was totally confused and wanted no part of the stint. He

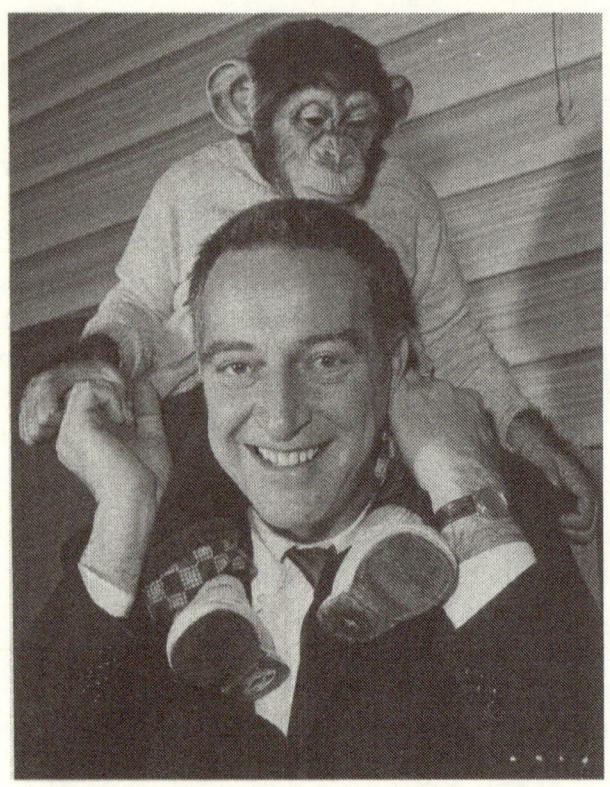

Zippy was one of Garry Moore's favorites

whined and fussed and wanted to go back to his dressing room. Who would have guessed? It turned out to be one of the most difficult shows we ever did! So simple and yet so difficult! We truly earned our pay that night.

Steve Allen loved having Zippy on his show. One night Zippy sat with Steve at Steve's famous desk, when Steve turned to Zippy and said, "You know, Zippy, you should have a show of your own." Unfortunately, that never happened. Edie Gormet and Steve Lawrence were the two most famous singers to get their start on the Steve Allen show and were married shortly afterwards. They invited Zippy to their apartment, where Zippy had a wonderful time. They both loved playing with Zippy. Later we went on the road with Edie for a short time through New England and into Eastern-most Canada.

Another memorable show we did on several occasions was "What's My Line." Zippy was trained to shine shoes for one of those shows and he thought it was great sport. He picked up the man's foot, placed it across his leg, pretended to apply shoe polish and then he buffed the shoe with a cloth. We had plenty of time for rehearsal, only with stand-ins and not the real stars, of course.

During the live show, the stars were blindfolded. Zippy came on stage carrying his little shoe shining kit and wearing a cobbler's apron. He picked up Henry Morgan's leg and as he exposed the bottom of Henry Morgan's shoe to the camera, the audience was treated to the view of a huge hole in the bottom of Henry's shoe. Henry's embarrassment was obvious, but the audience, cameramen and directors rolled with laughter. This show was one of television's best.

For many years we were contracted to work at an amusement park by a dear, wonderful agent named Abe Feinberg. Once a year, we picked him up in New York City and he would sit in the back seat of the Caddy with Zippy, as we drove to Dorney Park in Pennsylvania. Abe was famous for saying, "If God is willing." The same night we would return him to his place in New York City and not see him again until the next year. Zippy loved working at Dorney Park because they always let him ride the rides. Since I enjoyed them also, I would accompany Zippy. Although we usually rode them alone, everyone else would stand around and watch. We loved the Ferris wheel, but Zippy loved the roller coaster best of all. We sat together in the front seat and hung on for dear life. Of course, chimps swing around in trees in their natural habitat, so they have no real fear of heights. What a wonderfully fun time that was! Thank you Dorney Park for the years of enjoyment!

# Other Couples Who Worked with Zippy

Ralph and Kathleen Quinlan had a great skating act. As mentioned earlier, they lived next door to Zippy when he first came up from New Orleans to join the Howdy Doody Show. A baby chimp had been purchased and Kathleen absolutely fell in love. The little chimp wasn't getting much attention, so she decided to change that, and started training him. Because the Quinlans were trick skaters, they concentrated on teaching this new little Zippy how to skate really well. They ordered professional skates designed to clip onto high-top shoes that would support the chimp's ankles. The children's skates, with which he had begun his training, were soon found in the attic. Of the many couples hired to work with the chimps, Ralph and Kathleen Quinlan were the first, and they also stayed with us the longest.

Maybe it's just because I know the chimps so well, that I think this story is so funny. We used a highchair in our acts, as a home plate so to speak. It was a place for Zippy to return to as needed, show off his wonderful table manners, have his hula outfit taken off and on, take off or put on his skates, shoot the kids with his water gun, play his little toy piano and provide better visibility in general for the audiences.

This chimp had a tremendous amount of personality and a mind of his own. For a few days he played the game of "stupid", otherwise called "going blank on purpose", but better known to animal handlers as "testing". While Zippy sat on the arm of the chair, Ralph stood behind him. With both arms around Zippy, Ralph's hands were busy holding the piano in front of Zippy. Ralph's hands were occupied; his head a foot or two above Zippy's, when Zippy decided he didn't know he was supposed to strike the piano keys. Knowing completely that Ralph was in a helpless state, Zippy poised his index finger high above the keyboard and played dumb (instead of the piano). I have had this happen, and I promise you that there is no more helpless feeling than to be out on stage in front of thousands of people and have the chimp freeze.

After 3 shows of repeating this behavior, Ralph lost his patience. He placed his mouth on Zippy's ear and gently bit it. The reaction was simply wild! Zippy grabbed his ear, jumped out of the high chair, rolled on the ground into a crouch, first screaming and then barking in anger at Ralph, while still holding his "injured" ear. I doubt that anyone in the audience knew at that particular moment exactly what had happened, but they roared with laughter, and Ralph probably experienced the worst embarrassment of his life. For all you animal lovers please know that a chimp's ear is as tough as an elephant's, and he was not physically hurt at all. Zippy had simply lost the game of "stupid". From that point forward he would look up at Ralph and dutifully play the piano.

Ralph went on to become a pilot and bought a Cessna. Soon Zippy was flying high and in style. It was most impressive for Zippy to fly in for personal appearances.

When Ralph and Kathleen left, it was prudent to hire and train other couples to work with the chimps. Since I was in charge of training most of the chimps, it was natural for me to train the couples as well. It was an awesome job to: 1-train the chimps; 2-teach the act to the couples; and 3-train the couples and chimps to work together. Knowing the show, remembering what cue to give the chimp and when to give it, knowing what prop to use and when, knowing where the audience is and presenting the act to it, while maintaining a smile and control of the chimp and not tripping over a TV camera cable or falling off the stage…. This type of coordination is hard to teach and next to impossible when an animal is involved. Not all of the people had good stage presence or a good rapport with the chimp (or with each other, for that matter). They had to at least _appear_ to be a happy, united and a professional "family of three". Very few couples could measure up and there were many who stayed for only a short time.

The mansion of a home we lived in at that time, housed the couples quite well. After road expenses and agents' fees, they kept the larger portion of the money they earned working with the chimps.

We had a huge training room where chimps and couples were trained and rehearsals took place. A sewing room was necessary for making costumes. I was pretty good at that. An Olympic sized swimming pool was completely fenced in so that couples and chimps could play together in safety. The people could socialize and share ideas and war stories. Often the girls helped with housework or preparing a meal (for nine or more), and I was happy to relinquish those jobs at any opportunity.

As you might imagine, grocery shopping was difficult. Between the cases of fruits, vegetables and such, purchased for the chimps, and groceries for so many people, it was a full day's work. However, I have always loved to cook, and fortunately, most have liked my cooking. I was able to teach many of the girls to make specialty meals, and the kitchen was always a happy place to be.

One couple that came on board, Debbie and Joe Buglisi, were excellent in all respects. They were very good looking, handled their chimp well, and worked hard getting bookings, visiting agents, and creating new materials. They were often seen on the Merv Griffin Show. It was truly pleasant and comfortable to spend time with them. When they left, we called their chimp Debbie. Yes, the chimp was a female, and because of their loving way with her, she was just as loving. It was a sad day when they left us, but they had an exciting future ahead of them and we remained good friends for years.

When we first bought the chimp known as Chinners, the "Debbie" chimp became like a mother to him. They were so precious together! Debbie carried that adorable little chimp every place she went. She really spoiled me too, because she took a lot of responsibility off of me with her chimp-sitting. Little Chinners always walked around with his tiny little chin stuck up in the air. You might say that he led with his chin. Thus, the name Chinners stuck.

David and Holly were a great couple who worked with the chimps and we had much fun with them. Holly was a sweet person and quite a beautiful woman. Her beautiful blonde hair and delicate features tempted you to look at her and forget the chimp. David was a prince! Like Holly, he was good looking but was also the kind of guy that could practically read your mind and do things even before being asked. Both of them really loved their chimp, and you could see it in their performances. I spoke with Holly recently and learned she had been in a terrible automobile accident. She was literally run over by her own car. I still shed tears thinking about her.

Jack and Fran worked for us for a long time. Fran developed into an amazing trainer in her own right. She taught her Zippy, (Chinners), outstanding behaviors, including sign language that made you feel the chimp was human by the way he "verbalized" his wants. Also, her Zippy could tie his shoes and button his clothes, a feat most difficult for a chimp because of their small thumbs and long fingers. One night, Jack was booked on the Johnny Carson Show. At the beginning of rehearsal, Zippy put his mouth on Johnny's face, which is a customary greeting from an affectionate chimp. During the show, Zippy tied his shoes, buttoned his shirt and did many astounding feats for the audience, but Johnny ignored all that and kept talking about the chimp putting his mouth on him. What a waste! (There is more about this story in the forward). Sometimes the abilities of the chimps were so natural looking that people took for granted the time and training behind it all. Bad publicity like this, coupled with such facts as; liability insurance had become exorbitant, the price of chimps grew unbelievably high, government requirements and restrictions were getting tougher every day and animal activists were troublesome and became the death knell for many trained animal acts. Some time after my departure, Zippy Enterprises was basically no more.

After I married Don Womack, Jack, Fran and Chinners visited with us several times. We were constantly amazed by what that chimp could do. I left during the time Jack and Fran were at their best and never got to know any of the handlers after them. I was on a new venture, training exotic birds, and was out of touch. (There will be more about Alfie Cockatoo in the future).

# Zippy II

There are probably dozens of stories about this chimp, but I only know of a few. This is the chimp purchased just after Lee and Bonnie had arrived in New York City from New Orleans. Ralph and Kathleen trained and worked with him, and it was this chimp that greeted me at the door when I was first hired.

Here is a story about him that Kathleen liked to tell. Chimps will always let you know, with their body language, when they are ready for bed. While they were on the road, Zippy always slept next to their bed. Even chimps in captivity love to make a nest for sleeping. So their chimp was given towels, soft pillows, stuffed dolls and even the shirt he had worn that day to make a nest. Carefully, the chimp would tuck these things in around himself, rearranging each item meticulously until it met his satisfaction. Sometimes this ritual would take 5 minutes or more. Then the chimp would curl up in the middle of his nest, make the good night guttural "ooh, ooh" sound and fall into a deep sleep.

In those days, full skirts with lots of crinolines underneath were fashionable. They were bulky and took up a lot of closet space, so we would often hang them from a shower rod in the hotel room. Kathleen was proud of her collection and had crinolines in many different colors. One night, Zippy woke up, went into the bathroom and decided to clean the bathroom for Kathleen. He pulled down her black crinoline, dipped it into the toilet and began washing the bathroom walls. Over and over he repeated this action, into the toilet and onto the walls. Back and forth he diligently scrubbed away. Suddenly, Kathleen missed his presence, woke up and found him gone. She saw the light on in the bathroom, (yes, Zippys were well acquainted with light switches of all kinds), and found to her horror the bathroom had been painted black with the dye from her garment. Just imagine the Quinlan's embarrassment at having to tell the hotel manager that the bathroom was now black!

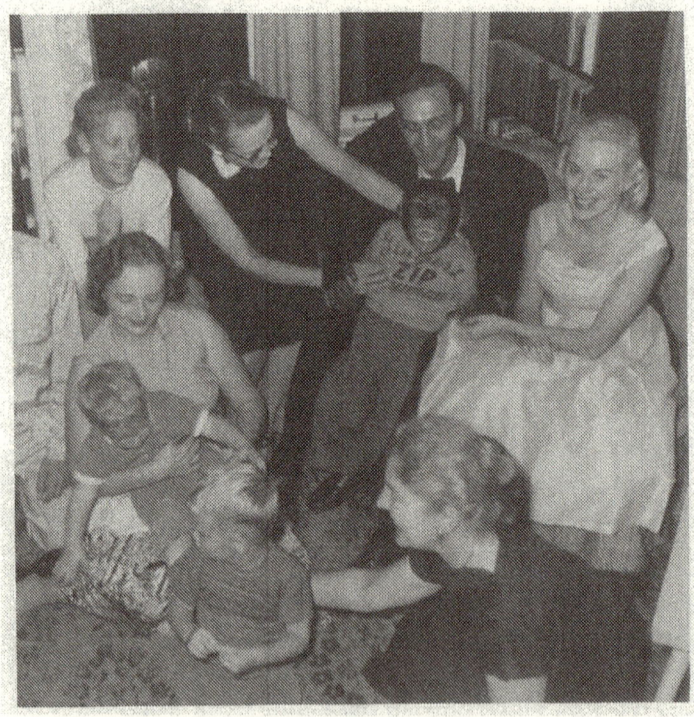

Zip and The Quinlans with some of
Carole's family

Ralph and Kathleen often worked sports shows. Usually there would be a tank of water placed in front of the stage for fishing or other water demonstrations or for use by the Lumberjack shows. Part of their act with Zippy was a jump rope routine which found Ralph and Kathleen swinging the rope for Zippy, as he would jump faster and faster. Then Zippy would make a lunge for Kathleen's shoe. Kathleen would drop the rope, and Zippy would chase her all over the stage, until he succeeded in getting one of her shoes. Then he would run to the front of the stage and throw it into the water, or if no water tank was there, into the audience. It never failed to bring the house down. No other chimp ever did this with the diligence and persistence of Zippy II. It could just never be duplicated no matter how hard we tried.

Kathleen became very ill as her chimp came to the end of his season. Their escapades together came to an end, as did a beautiful, long-term, special friendship. Kathleen was a special kind of woman who loved man and beast with all her heart.

# Lil'Shaver, the Pouter

When I arrived on the scene, there were two Zippys, Snow and Lil'Shaver, a total of four chimps. You see, there were always two Zippys and another one in training. Snow was a white chimpanzee, and Lil'Shaver went on to become one of the Zippys. He was the very first chimp I ever personally trained; my first little prodigy. I was given total leave to train him as I saw fit, and I poured my heart and soul into him.

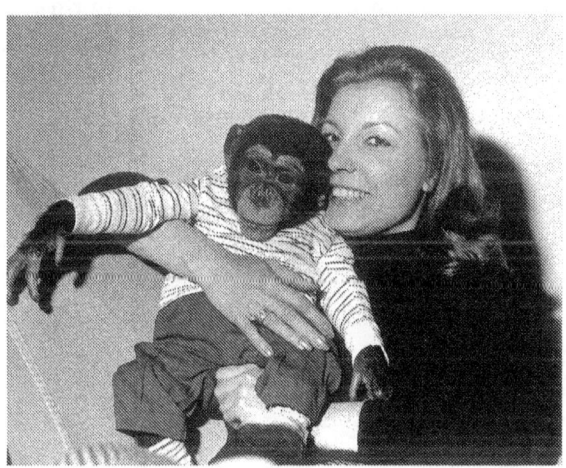

Carole holding one-year old Lil' Shaver

So much was going on with the two Zippys, that Snow and Shaver received almost no attention, until I came along that is. Snow has a chapter of his/her own, but Lil'Shaver's is a precious story I will share with you now. I pray that the Lord above has a special place for all these little guys; because someday in heaven I want to hold, squeeze and converse with these little loves again, especially my Lil'Shaver.

Shaver always seemed to be a sad little chimp. He had a habit of moving his mouth and lips in a way that made him appear to be pouting and downhearted. Fun was not an important part of his life. Being with me was all he cared about. He was about a year old when he arrived, and later he adopted me as his mother. Nothing would deter him from me. After working with the disappearing chimp, (the original Zippy), it was a joy to have one that was not bounding off all the time. Unfortunately, he didn't have quite as much personality, but he made up for it with absolute obedience to me. While my approval was important to him, Shaver and his daddy had no rapport. Now, it always *appeared* on stage as though he was obeying his daddy, but he really was not. He would glance over at me to see if he was to obey the instructions. It was mostly my body language that told him to listen and do what he was told to do. Again, this chimp was fabulous for TV. He need only be shown once what to do, and he did it. However, his personal appearances were not as good. Shaver had less spark or devilment on the stage, and his routine was just that, a routine, even though he did it without a flaw. There was less mischief or spontaneity to it, but there was almost nothing I couldn't teach him. So he appeared on numerous television shows and made many commercials.

Shaver was the chimp used in the Castle films. Some of the feats done in those films were complex and fascinating to those who understand animal training. The films, "Zippy's Beach Adventure" and "Zippy's Birthday Party" feature Zippy, (Lil'Shaver), his daddy and the blonde, (me). The "Beach Adventure" film was taken on Lido Beach in NY, and the temperature was in the 90's. What a trooper that little guy was! The "Birthday Party" adventure was taken in our home. If you want to see an example of the pouting look I mentioned, check out those two films.

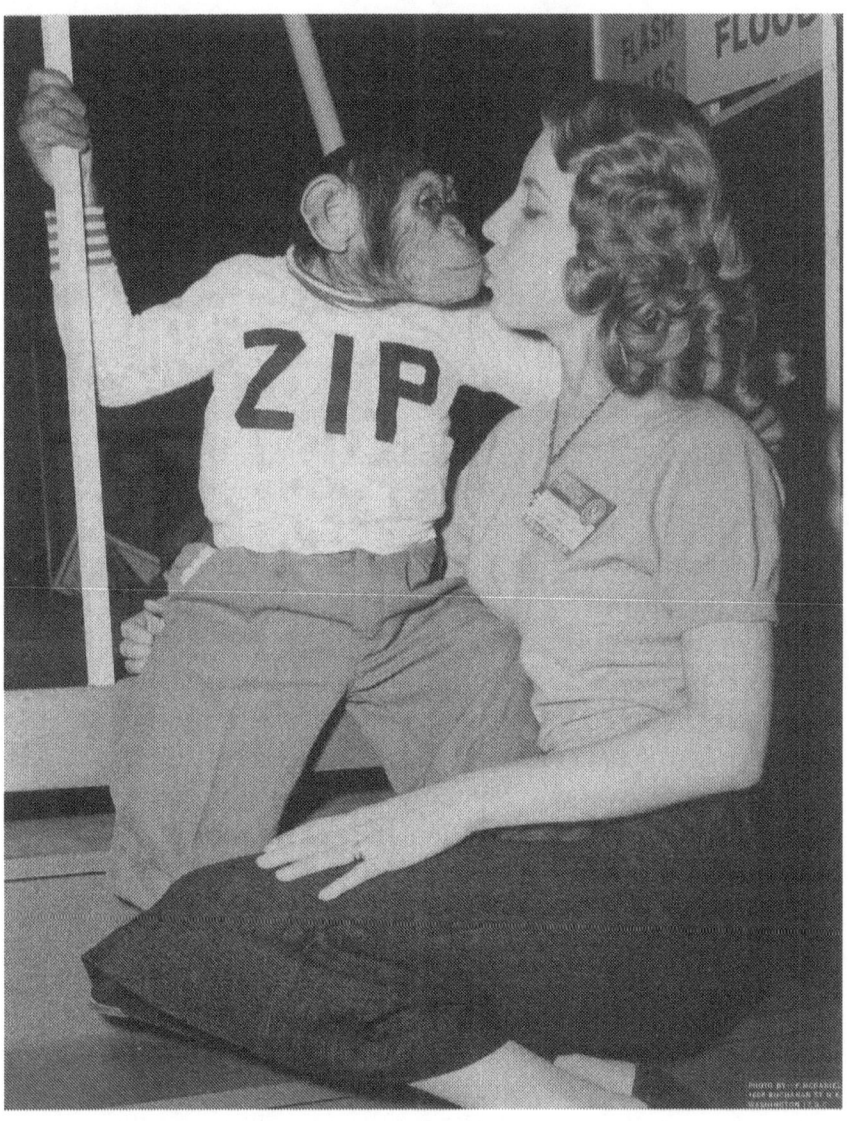

Zippy sneaks a kiss at a trade show in New York

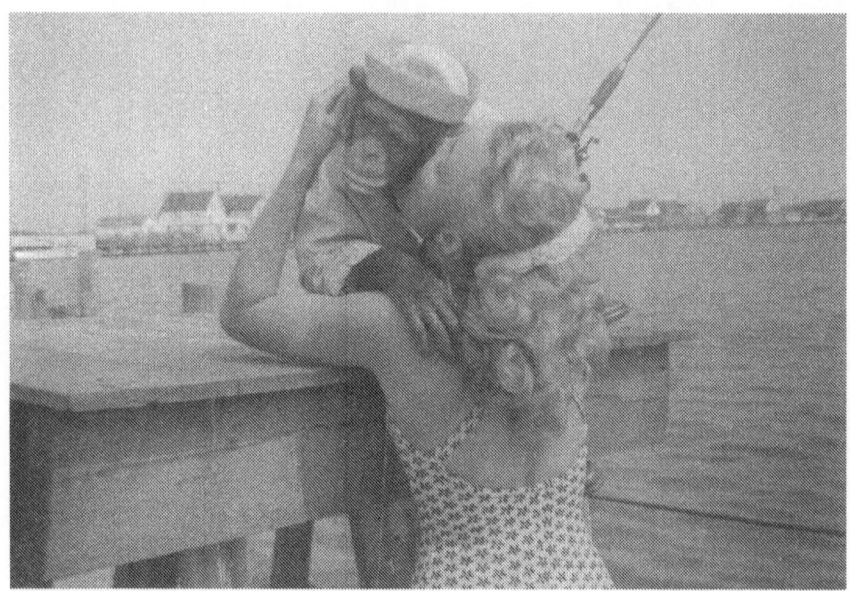

Zippy sneaks another kiss during a break from shooting
Beach Adventure for Castle Films

I have so many wonderful stories to tell about the little pouter. Because he was so docile, we could go everywhere with him. When we were on the road, we would stay in hotels and motels. It could be a boring time for him, and often we would take walks with Zippy/Shaver to pass the time and get a little exercise. It was part of our standard procedure, to bring along a stroller. We would wrap him up in his little coat and cap. He would jump into the stroller, and off we would walk down the streets of Anytown, USA. Most people would pass right by us and never notice we had a chimp in the stroller instead of a child.

Even if the chimp didn't have much mischief or "monkey business" in him, we certainly did! We had such fun when we would come to an intersection, especially if everyone had stopped to wait for the light to change. The stroller would come to a stop beside some unsuspecting individual. As we waited, Zippy, being the sweet little love that he was, would reach up and take hold of someone's hand. I assume the person had noticed a stroller in his or her peripheral vision, but had not looked closely enough to notice that a chimp was the passenger. Therefore, the person would casually accept the chimp's hand, but within seconds would sense that this was no ordinary child's hand, being a little too hairy and rough. Within a moment or two the person would look down, gasp, and pull his or her hand away. Then the person would look at us strangely as if to say, "What have you done to me?" We would immediately introduce Zippy, the famous chimp on television, and invite the stunned individual to whatever show or personal appearance we were doing on that particular day. Usually, they would show up, bringing friends and family to prove that their story was indeed legitimate. At that point, they would proudly introduce their folks to us, and of course, Zippy. However, things did not always go so smoothly. We had a smattering of screamers, overly dramatic teenage girls, prudish old ladies and overprotective mothers, who would invariably say, "Be careful, darling. He might bite you!" *Then* the poor child would immediately and unnecessarily become fearful.

One of my favorite things to do was to fall behind several yards and appear to be looking into a store window. As people passed, they would make comments that I could overhear. Mostly, it was "Isn't that a shame, such an ugly child!"

One time I was sitting with Zippy in the front seat of a car. Two women were walking together and talking, while a little girl followed them a few feet behind. The child spotted Zippy and ran up to her mother. She grabbed her mom's hand with her little hand and pointed to Zippy with the other. As she tugged on her mother's hand, she was shouting, "Momma! Momma! There is a monkey in that car!" The mother took only a quick glimpse. Obviously, she was thinking it was a deformed child, and she tried to hush her child and drag her along. However, the little girl knew perfectly well what she had seen, and kept tugging insistently, "Momma! Momma! It _really_ is a monkey!" The mother tried hard to ignore her daughter and quiet the little one, so my supposedly deformed child and I wouldn't be embarrassed. They kept at this all the way to the corner, crossed the street and then crossed the other way.

All the way down the street, I could see the child tugging and pointing as her mother literally dragged her into the distance. The friend never once even turned her head. I wanted so much to take Zippy and run down the street after them. I knew it would be a great thrill for them all, but we were in a hurry to make a television appearance. I often wonder if that mother ever learned the truth. Not likely!

One of my favorite stories happened in a department store in Long Island, New York. We were fitting Zippy for a suit he needed for a big television show. As we were paying for it, Zippy was sitting on the back of a couch, watching the people that were rushing by him. Along came a mother and father with their two young children. The mother first spotted Zippy and was eager to point him out to her kids. She stopped until the children caught up and said, "Look, over there on the couch is a monkey." The little girl said, with great disdain in her voice, "Mother, that's not a monkey, that's a chimpanzee." Her comment was priceless and I wanted to hug that little girl. I had grown so tired of the standard, "Hey, Joe, there's your brother!" routine, and of course, she was right. A chimpanzee is a member of the ape family, not a monkey! However, we had learned to stand and smile and pretend that everyone is so original.

Laugh at their jokes and keep your mouth shut when someone says, "monkey". "Grit your teeth and take it, Carole. That's why you make the big bucks." I would say to myself.

Lil'Shaver was the easiest to manage of all the chimps. Traveling with him was a dream.

Usually we would go to a motel or hotel room, and I would take the pillows off the bed, tuck all the covers in, place Zippy in the middle and give him a pile of toys and pillows. He would play until he was sleepy, at which time, he would snuggle into the pillows and fall asleep.

Unfortunately, there were a couple of times when it didn't quite work out like that. The first was at the Utica Hotel in Utica, NY. We had gone through our usual preparations and then placed a chair for each of us on either side of the bed. In those days, room TV's were placed on gold-colored, light weight, little roll-around stands. We rolled the TV up to the foot of the bed and sat down to watch a Tarzan movie, thinking nothing could go wrong since we were so close to the bed. Right! As the suspense built, Tarzan walked across a stream on a fallen log. Cheetah, (the chimp), gingerly tried to follow Tarzan and seemed very scared, and for good reason. First of all, chimps cannot swim; but even worse, several crocodiles were in the water snapping at Cheetah. Cheetah was screaming the fear sound that chimps make. Although we knew it was only a movie, it appeared to Zippy to be very real. He stood up in the middle of the bed, began to jump up and down and scratch his sides, (exactly as movies and humans portray). He started the "whoo, whoo, whoo" sounds, (which build and progress to a terrifying scream), as the hair on his body stood straight out. Suddenly, Zippy leaped in one bound to the foot of the bed and hit the nearest crocodile as he let out his final scream. He sent that crocodile, Cheetah, the television and stand rolling across the room, crashing against the wall. Then he quietly sat down to play, satisfied that he had defended his little friend on TV and killed the offending, dangerous, monstrous crocodile. Naturally, we were quite worried about what damage he might have done, but I'm happy to report that the crocodile, Cheetah and the television went on to live another day. Of course, we will never tell Zippy, the unsung hero.

Scranton, Pennsylvania was another location where things didn't quite go as planned. We were booked to perform at a fair that summer and we were staying at a nearby motel. The temperature was in the 90's, and there was no air conditioning. We were basically using the motel room as a dressing room and a place to let Zippy rest between shows. The bed was prepared as usual, and Zippy was happily playing with his toys. I was in the bathroom getting dressed. I had just put on my one-piece undergarment and started hooking my stockings, when I heard the screen door slam shut. It was too early for Zippy's daddy to have returned, so I peeked around the door and viewed an empty bed. That could only mean that Zippy had left the room. It was very out of character for this chimp, but there was no other explanation.

The nearest thing to me was my full circle skirt, so I wrapped it around me as I took off after the chimp. Motels are located on busy streets, so I was very worried. There wasn't a second to lose! I went through the doorway and came to a sudden stop. Zippy was nowhere to be seen. During the day, in the parking lot of a motel, there are seldom any cars for him to use as a hiding place. He couldn't have gone far in just a few seconds. Where was he?

Suddenly, I heard laughter coming from two rooms down. I hitched the skirt around me and made my way to the door. As I looked in to survey the situation, I saw several men seated around a low round table. Zippy was going around the table taking each man's glass, drinking the contents and placing the glass neatly back where he had gotten it. I tried to call him to me; but he ignored me without so much as a glance, and proceeded around the table, helping himself to their drinks. The men thought this was terrific fun and continued to refill the glasses for the chimp. I was invited in, but said, "No, thank you. I'm not exactly dressed for the occasion. Could you hand the chimp over to me, please?" Not one of them made the slightest effort to help me in my predicament. They explained to me that they were Dr. Pepper salesmen, but still made no attempt to deliver the chimp to me. I was frustrated but also relieved to know that Zippy had been sampling soda, rather than liquor, which I had suspected was in the drinks. Eventually I was able to coax the chimp over to me, probably because he had finally drunk his fill of the soda. The men had enjoyed the visit tremendously, but that's not where the story ends.

Another man had just parked his car in front of the doorway, and he was starting to get out when he spotted me. You see, in my haste to find Zippy, modesty took second place. Although I was partly dressed (and definitely <u>not</u> naked), I could imagine how the situation looked to the new arrival behind me. I will never know how long he was standing there ogling. However, when I turned to leave, with Zippy holding my hand, his eyes got even bigger and his mouth hung completely open. He stood in disbelief as this half-naked woman and her chimpanzee passed by him and disappeared into another motel room.

For a chimp that rarely got into mischief, he really outdid himself that time. He couldn't have planned a better practical joke on his mommy if he'd tried. You have to learn to laugh at yourself sometimes. The biggest embarrassments can lead to some of the most humorous memories with the passage of time.

# Snow, the White Chimp

Snow was a startlingly beautiful chimp, with a face that was almost pretty. He had bright blue eyelids, very blonde, (almost white), hair, brown eyes, and fingernails with moons, like humans have. In fact, he was so pretty that we dressed him as a little girl and always referred to him as female. Henceforth, I will usually refer to Snow as "her".

She was definitely not an albino, but no one could explain why she was blonde. She was simply a freak of nature. Some people even went so far as to suspect there was a human in the monkey barrel, but scientists disputed that theory. She was just different and beautifully so.

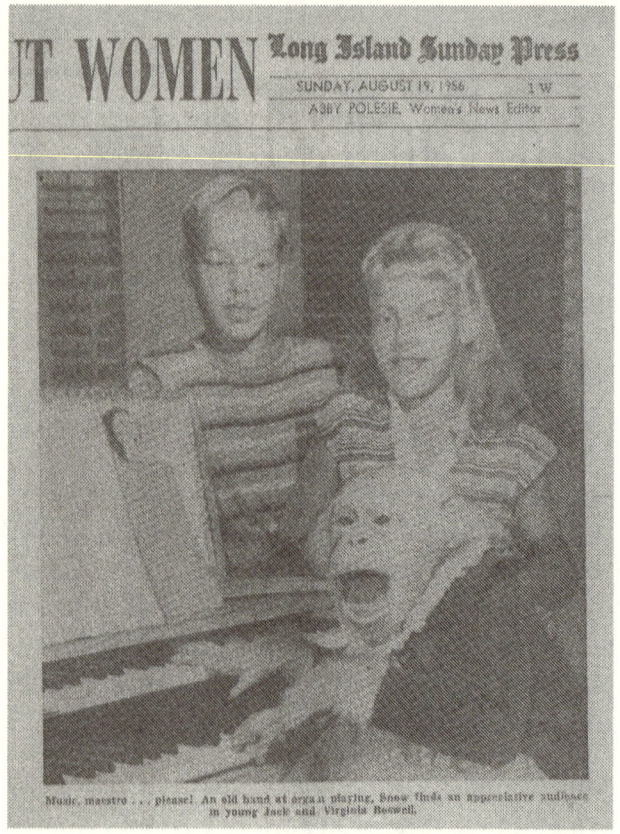

Famous white chimp accompanies herself at the organ

Unfortunately, she never drew the attention that we had hoped she might. Zippy's daddy paid $12,000. for her, thinking she would be a terrific showpiece, but it seemed most people were unable to grasp the concept that she was "the only white chimp in the world." When I would tell people specifically, "she *is* the only white chimp in the world," the standard reply would be, "Come to think of it, I've not seen many of them." Somehow it just didn't compute.

There was a Believe it or Not type museum on West 42nd Street in New York, just down from Times Square that featured pictures and articles about Snow. Once, when in Florida, we went to the "Believe It or Not" museum there and saw pictures and articles about her.

The first day that Snow had arrived at the house, she spotted a raw roast on the counter in the kitchen. She grabbed it, ran to the corner and devoured it. She was very aggressive and would not let anyone take it from her. Snow had growled and threatened everyone and consequently, everyone was intimidated by her, after seeing this aggressive side to her personality. No one told me this until years later. Therefore, I never thought to be afraid of her and got on very well with her. I met Snow my second day on the job, had dinner with her, as a matter of fact. I liked her instantly, but then I was enamored with all the chimps at that time. There was an affinity there, with the chimps that was to last for the next twenty-three years.

Snow loved to play and tumble, especially tumble. It was fun to dress her in her pretty dresses, and she loved her clothes and treated them well. Because she tumbled so much, I bought her ruffled little panties so that when she was upside down, her little bottom would be attractive.

Over the years, I taught her many tricks, but she never quite caught on to the idea of skating. We began with clip-on skates, and she never advanced to the professional skates that the other chimps used. Snow would walk and run and jump on her skates, but she never could understand how to push them and glide on them. However, she was so funny on the skates that the laughs made up for her lack of talent.

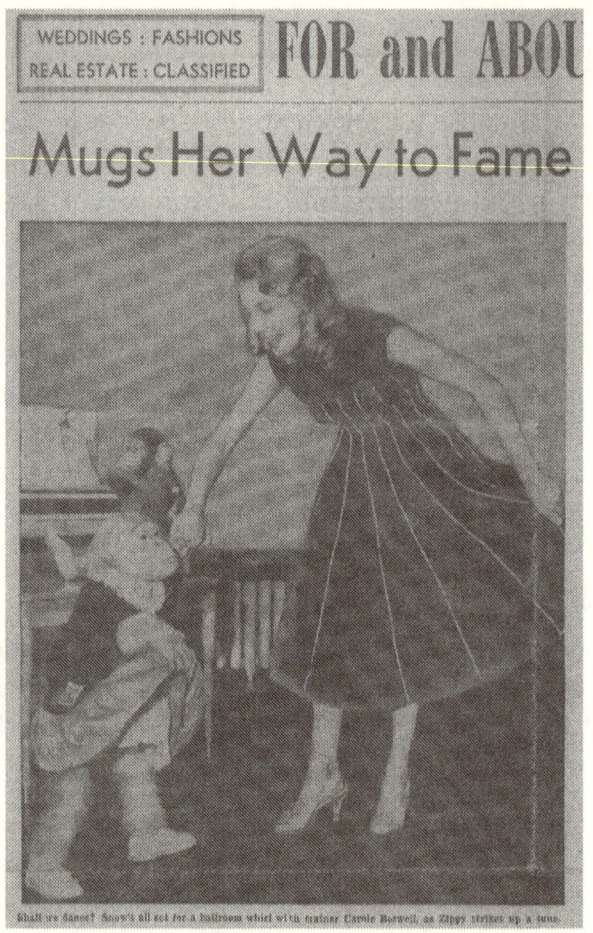

WEDDINGS : FASHIONS
REAL ESTATE : CLASSIFIED
FOR and ABOU

## Mugs Her Way to Fame

Shall we dance? Snow's all set for a ballroom whirl with trainer Carole Boswell, as Zippy strikes up a tune.

Ballroom dancing with Snow and Carole

We tried very hard to find a place for Snow, but nothing ever seemed to work out. She made a few television appearances. Gary Moore used her a few times, but preferred Zippy, who was on regularly. Arthur Godfrey used her on his show once. He was very impressed with Snow, and she loved tumbling in his lap. Arthur Godfrey invited us to his home in Virginia. We had lunch with him, and he gave us a grand tour of his beautiful estate. One of the things I remembered and impressed me the most was a TV set that was hung from the ceiling on four chains above the foot of his bed. I thought about how he could lie in bed and watch television. How neat! Arthur wound up asking us if Snow could be a permanent member of his show. At the time, we were under contract with Zippy to CBS. So, we told him yes, but that our agent with CBS would have to be involved, and that we would have him contact Mr. Godfrey for a meeting. We never heard from Mr. Godfrey again. Later, we were told that he liked to have more of a personal relationship with the folks on his show, and that he had been hurt because we were so formal about the business end of it. There may have been more to it than that, but at this point, I will probably never know for certain.

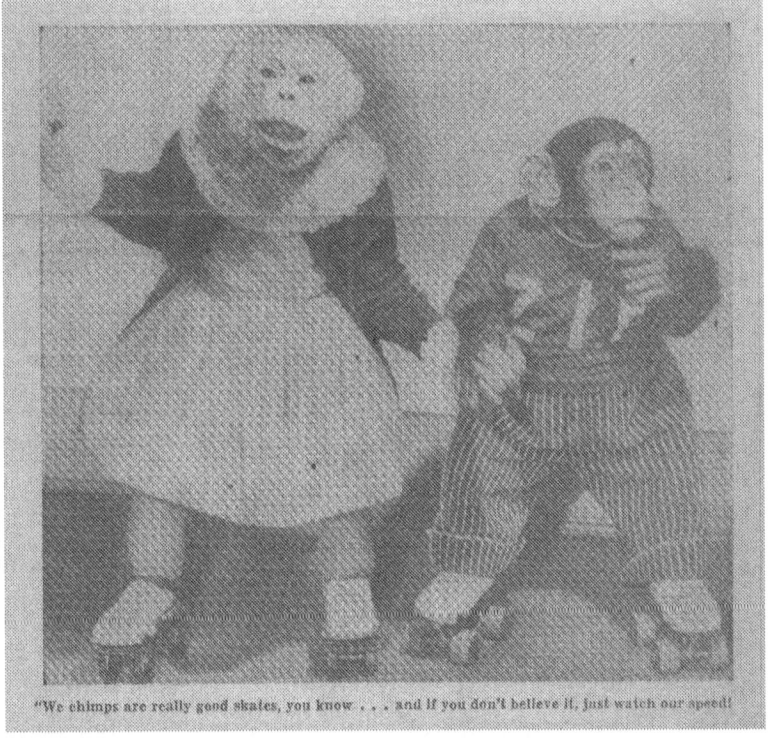

"We chimps are really good skates, you know . . . and if you don't believe it, just watch our speed!

Skating Snow and Zippy (Lil' Shaver)

Perry Como had Snow on his show once. I know the show went well, but I don't remember it, because Ronald Reagan was also on the show the same night. My eyes and attention were on him. He has always been special in my eyes, and I was so impressed with him. We talked about his daughter, Pat. Her birthday was coming up soon, and I offered to send her a Snow doll for her birthday, which I did. Pat sent back some candy for Snow.

The Snow doll was produced by Gund Toy Company, I believe. Although it was an extremely lovely doll, it didn't stay in their line for very long. I would love to have one now and I often search eBay to see if a plush Snow doll will appear for sale.

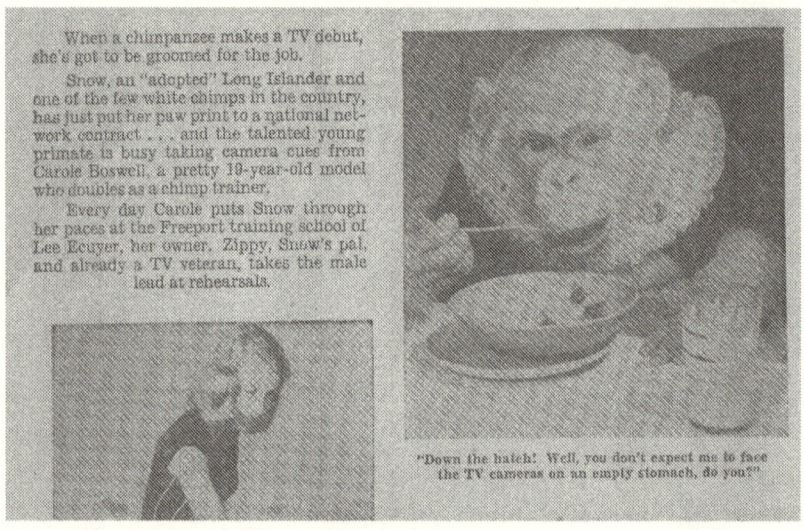

When a chimpanzee makes a TV debut, she's got to be groomed for the job.

Snow, an "adopted" Long Islander and one of the few white chimps in the country, has just put her paw print to a national network contract . . . and the talented young primate is busy taking camera cues from Carole Boswell, a pretty 19-year-old model who doubles as a chimp trainer.

Every day Carole puts Snow through her paces at the Freeport training school of Lee Ecuyer, her owner. Zippy, Snow's pal, and already a TV veteran, takes the male lead at rehearsals.

"Down the hatch! Well, you don't expect me to face the TV cameras on an empty stomach, do you?"

Impeccable table manners

As time went by, Snow grew and matured. When he reached sexual maturity, it became necessary to find a home for him. The Philadelphia Zoo offered him a home, and I believe they kept him for about a year. Unfortunately, they had problems with him. The chimp battered the steel door so badly that they couldn't open it at one time. He also masturbated, quite often in front of the viewers and was a cause of much embarrassment. Eventually, the keeper of the zoo called us and told us there was a research center that wanted to try to breed Snow, in an attempt to produce more white chimps. We agreed, but only after examining the laboratory and assuring ourselves that this would be a fit and proper place for our friend. A year or so later, we were told that Snow was sterile, and all attempts to mate him had failed. I didn't want to know more about him after that, but I'll never forget him. Snow was one of those challenges I am proud to say I mastered, at least for a while.

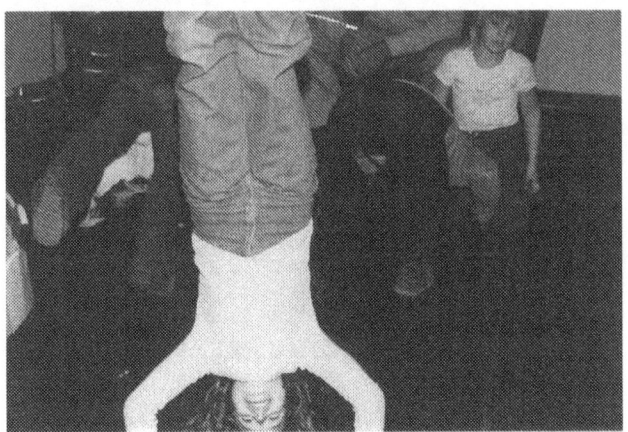

Practice makes for perfect headstands

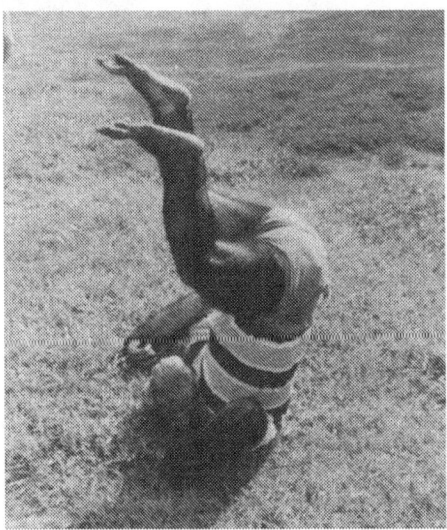

I've been practicing, but I think I need a flat head

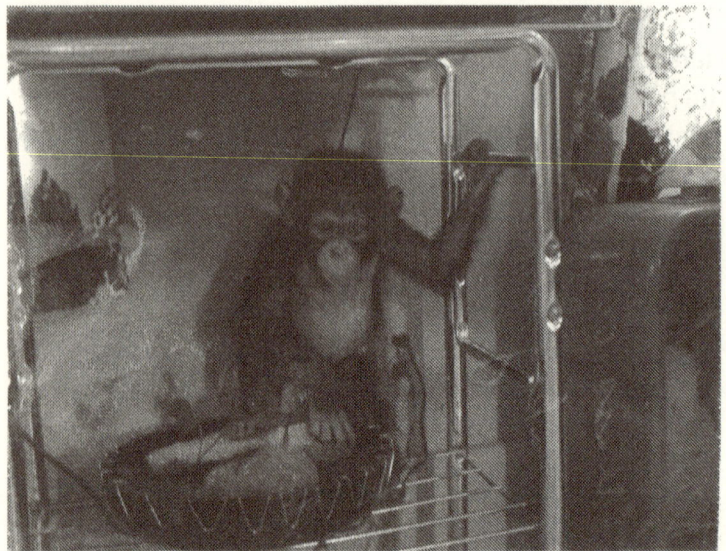

I don't think this is a Jungle Gym

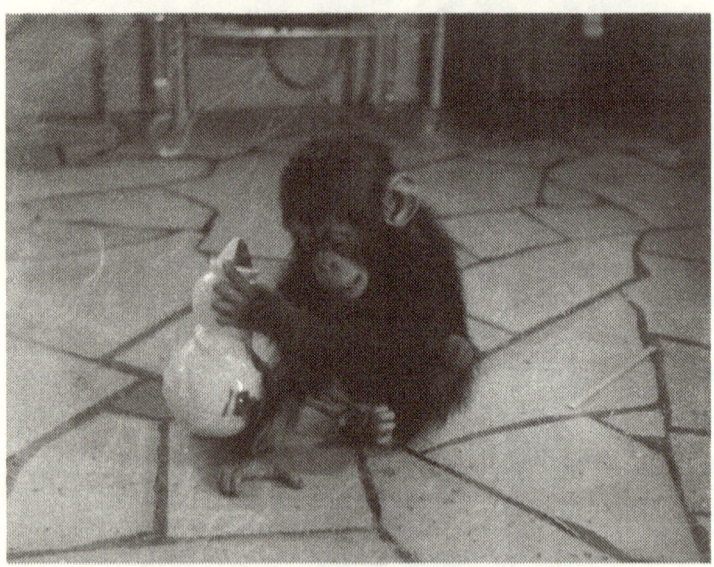

I never saw one of these in the jungle

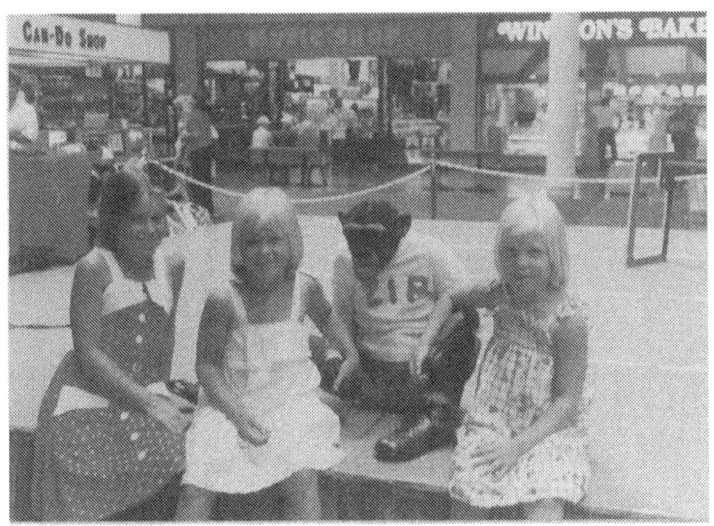

Our girls relax with Zippy between shows

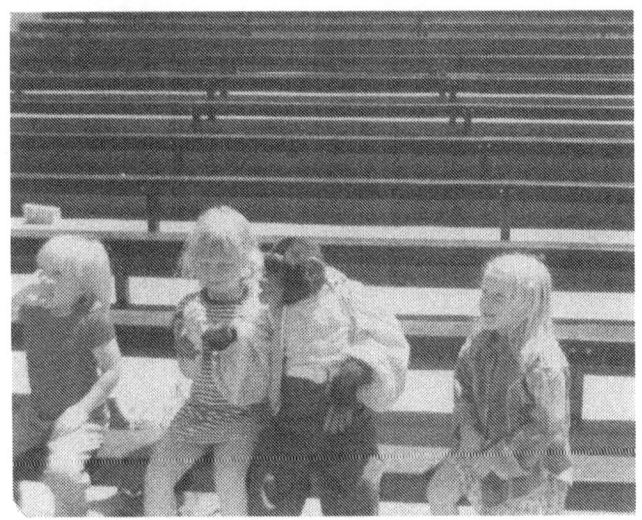

Big or small, the ladies love Zippy

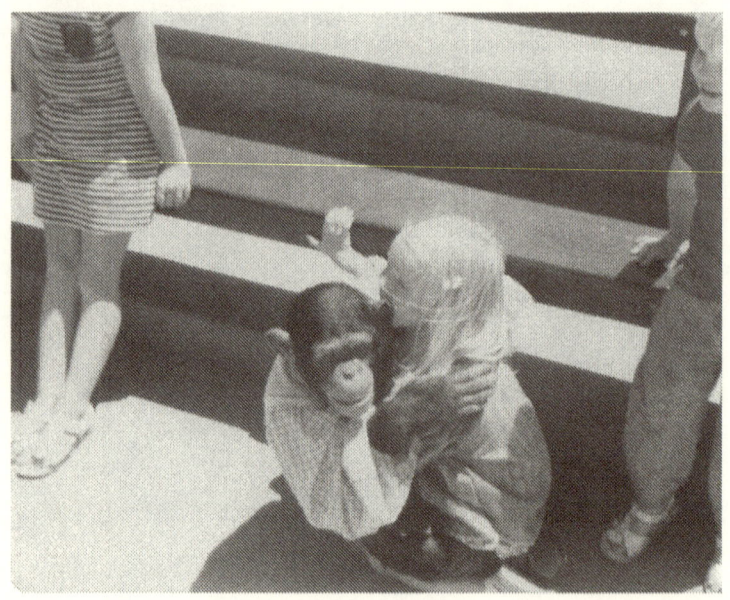

And Zippy loves them, too

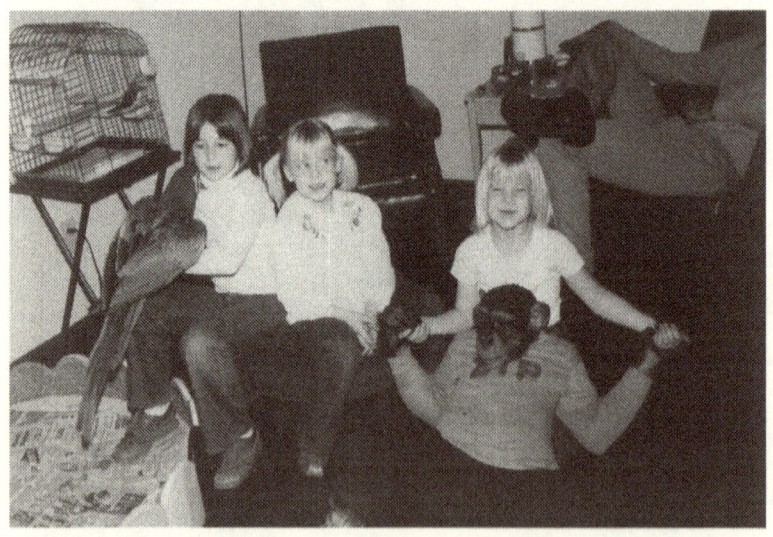

Our girls, Zippy and Mickey, our comedy macaw

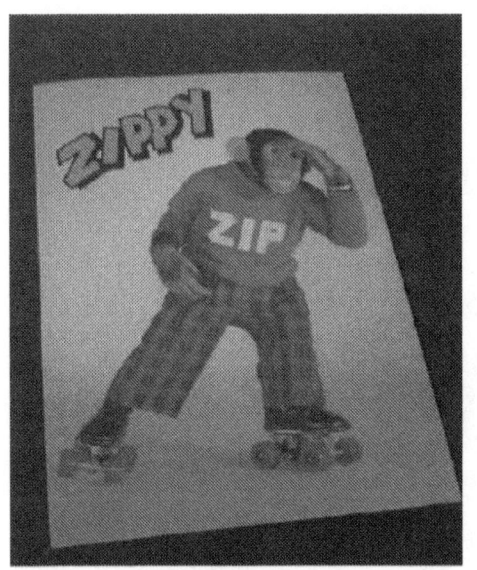

Where did everyone go? (Fan club card)

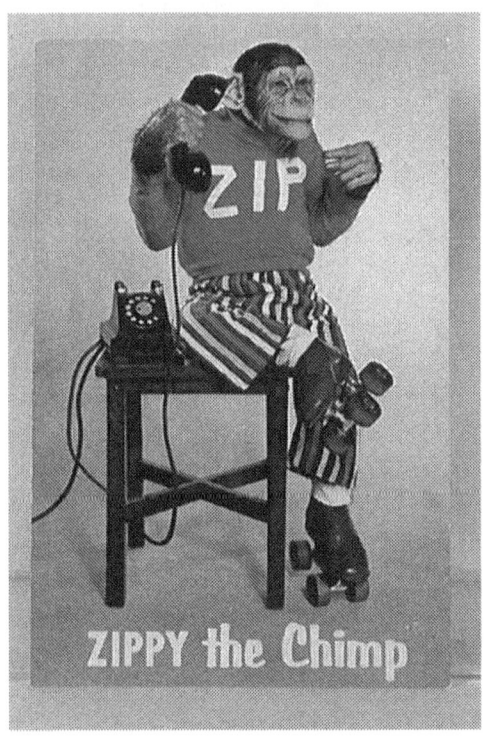

Quick, get my agent! (Fan club card)

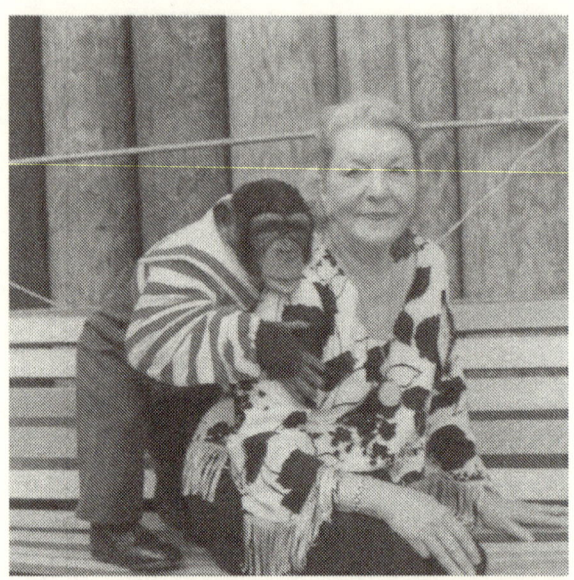

I love you, Grammy

And I love you, Don

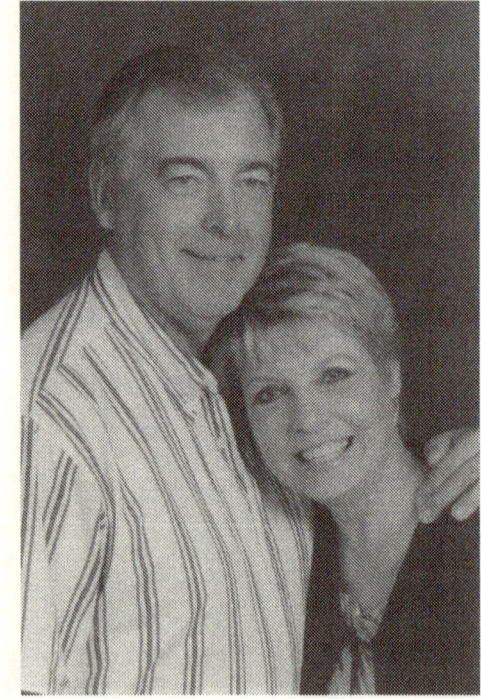

# How We Lived and Worked with Zippy

The chimps stayed in closed-up rooms, which basically served as their cages. I always wanted them to be near me and I spoiled the little ones as much as possible. So, I would choose one in the evening and give him a bath, (they all loved bath time), complete with toys, rubber ducks, the whole gamut. After the bath, the chimp was toweled dry, diapered, leashed, dressed and brought onto our king-sized bed. There was a large TV at the foot of the bed. Zippy and his toys were always in the middle, with his mom and dad on either side of him. Others in the household (trainer/performer couples) gathered around on chairs, with their feet up on the edge of the bed. Please understand; it was a huge bedroom, in a huge house.

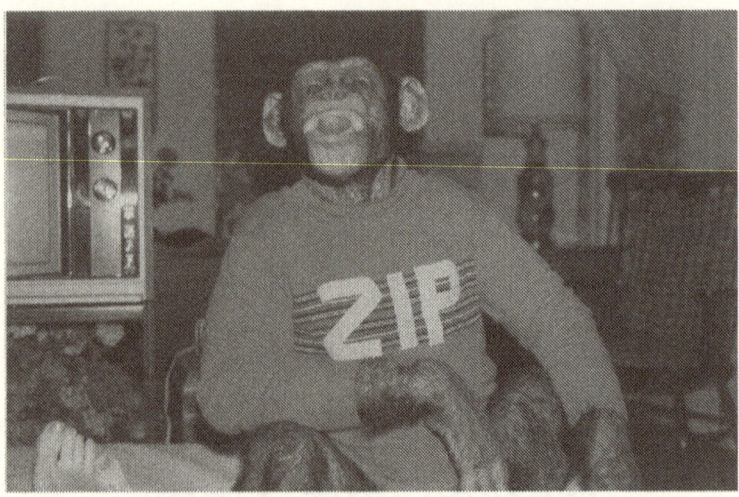

A sassy happy chimp

The chimps always loved to be on our bed when we traveled, too. It was the easiest way to relax and sort of "corral" Zippy. They quickly understood that this was conventional, comfortable and fun. When we went into hotels and motels, each Zippy would head for the bed and wait for his toys. So, in our home, this was normal and natural for us. Bowls of popcorn, fresh-made fudge and soft drinks were everywhere. We all had such fun, and it was good, clean, wholesome fun. After the 11:00 o'clock news, everyone, (including Zippy), returned to their respective bedrooms and all was well in the world.

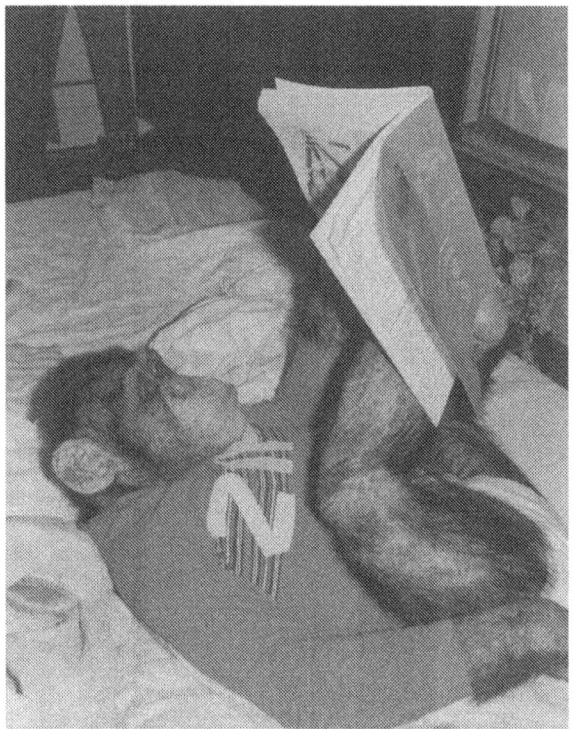

Reading with both feet in total comfort

I was always loving and hugging on the little guys. They were so much like small children. As I write about it now, the loving feelings well up in me. They loved me, hugged me, slept in my lap, ate at our table, rode in our vehicles, flew in planes with us, walked or skated down Broadway or Fifth Avenue, or skated across the street to play in the park. For me they were almost human. Health Departments were not so strict in those days, and we often ate out in New York or when on the road with Zippy. Scores of stories can be told about eating in restaurants with Zippy. I made sure that all the Zippys had good table manners. Napkins were tucked into their turtle-neck shirts; and they knew how to properly hold a fork or spoon, drink out of a cup or glass, and place it on the right, (above the knife, of course), etc. Hundreds of people would say that they wished their kids could be so well behaved. People have actually asked me if I could train their kids for them. In some cases, I believe they really meant it, and there were surely some that could benefit from a little training.

We could never get away with some of the things today that we did then. The Health Department would close down the restaurant, *if* we were even able to get into one in the first place. Animal activists and other groups would scream bloody murder for animal cruelty. "The very ideas of making a chimp wear clothes and shoes and walk upright! Those people are so cruel!" Can't you just hear them? Not to mention the fact that people today would sue if they were simply startled by the sight of a chimp. In fact, I will give you a couple of examples.

On one of the early days in television, we were in the NBC building for a show. Zippy hugged a woman there. She tried to sue under the pretense he had hugged her for 10 minutes, and she acquired a headache and subsequently had to take an aspirin. Lloyds of London contacted her and offered to pay for her aspirin. On another occasion, at a sport show, a model held her arms out near the stage where Zippy was rehearsing. This was Zippy's cue to jump into my arms, so he did what he thought he was supposed to do and jumped into *her* arms. Zippy's skates scraped her thighs as his feet went around her waist. She tried to sue, but nothing came of it. In *today's* legalistic and "sue-happy" society, she might have prevailed.

We ate with Zippy at some pretty ritzy restaurants. Zippy often frequented the famous Sardi's, but there is a special story about Zippy and an Italian hide-away called "Lonnie's." Some executives from CBS had invited us to lunch there. We were discussing our contract with CBS, and I wanted Zippy to be at his best. Lonnie always brought Zippy a dish of Italian bread and some sauce, so Zippy could dip/soak his bread. That day, we were all talking and I failed to notice that Zippy had run out of dip. For a while, he sat very patiently, but when he couldn't get my attention, he spotted the cup of coffee belonging to the man sitting opposite us. Picture Zippy making his calculations.... He carefully and accurately tossed the bread into the gentleman's cup, splashing coffee all over his tie and shirt. Fortunately, this man was more fascinated in Zippy's accomplishments than he was worried about his tie and shirt. Chimps are very clever.

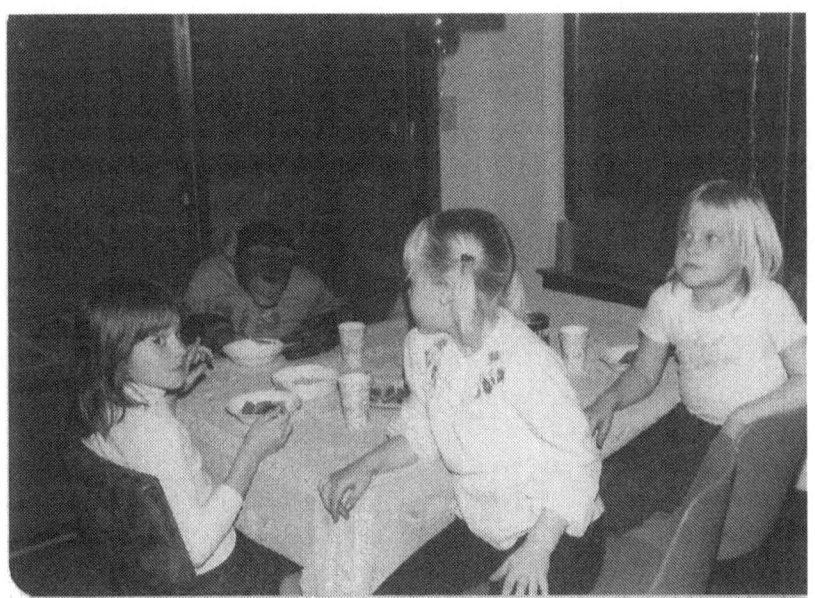

Our girls finish dining with Zippy

Surprisingly, we got the CBS contract, which I believe lasted for 4 years. The contract basically stated that if a star did not show up for a TV show, of one kind or another, Zippy was to fill in for the missing person. We were always on call and consequently were on more TV shows than I can remember. We often got a call at the last minute, so we had Zippy's simple props, bag of fresh clothes, and his famous roller skates ready and waiting at the back door, so we could scoot out in a hurry. We would get a call, and 45 minutes later, be at the studio, and Zippy would do his thing. I don't think there was ever a time that Zippy wasn't the hit of the show. It was uncanny how Zippy just knew he was a star and found ways to make the audience laugh. I can't remember when there was ever a problem. He always seemed to like everyone in a TV studio or on stage. It was as though he recognized his calling in life. We had hoped that CBS would write and produce a TV show around Zippy, but all that ever transpired along those lines, were two short-lived comic books.

We always let the Zippys be as free as possible, especially in public. We were known for Zippy's not requiring a leash, working without restraints and for the fact that any big star could work with Zippy *alone* on stage, while we directed from the sidelines. The Zippys I trained and worked with took direction well. Zippy was on hundreds of television shows where we were not seen, but were giving him his cues "behind the scenes". More often than not, the shows would be done with no rehearsal. Jackie Gleason *never* rehearsed. He had an understudy sit in for him, and he would only go on stage for the actual show. Many surprises occurred on that show, and others.

One time we did the Kraft Theatre without a real rehearsal. The show starred Hal Marsh. It was a South Pacific setting in a barracks with soldiers sitting on their bunks in their skivvies... casually dressed. During the play, Zippy was to walk through the barracks, do some simple stunt and walk out another door, where I would meet him. When no one was on the set, we showed him what to do. However, at show time, Zippy got distracted by one of the actors and made a detour no one expected. I carefully peeked around the doorway to see where he had gone and there was the camera, focused on me! Zippy had actually done the stunt correctly, but had turned so that his back was to the camera as he was heading toward the second door. I stood up and realized I was staring the cameraman in the face. The cameraman was so surprised that he didn't turn his camera off, or away from me. Suddenly I heard the director say very loudly, "How'd she get there?" He was very angry with me. I was in a cocktail dress and certainly didn't belong in *that* setting. Since the show was live, I never got to see it. I'll bet it was interesting. Needless to say, we never performed on the Kraft Theatre again.

Some shows used Zippy numerous times, such as Howdy Doody, Gary Moore, Jackie Gleason, Captain Kangaroo, Ed Sullivan, to name a few. Howdy Doody was a fun show to do. I was involved in their shows on the road. During that time, Princess Summer Fall Winter Spring, (or however her name went), was killed in an unrelated automobile accident. It was a terrible tragedy and frankly, she could never be replaced on the live TV show. I do not believe the producers ever tried to replace her. I was Princess Carole in the road show.

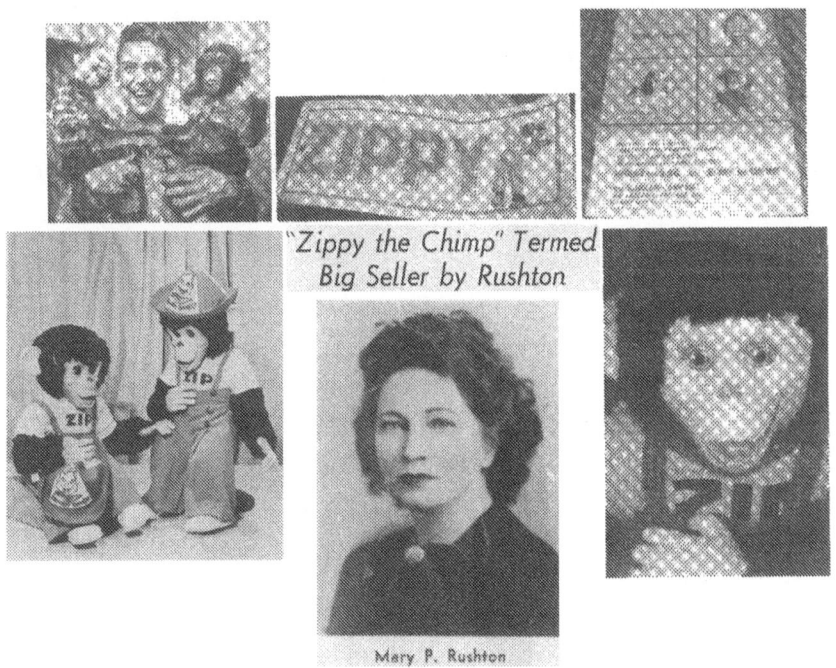

"Zippy the Chimp" Termed
Big Seller by Rushton

Mary P. Rushton

Bob Smith, Howdy Doody, Zip and the 1980's Zippy doll

There are two episodes that always come to mind when I think of the Howdy Doody Show. One is the time the little boy came up to Buffalo Bob and indicated that he needed to re-lieve himself. It was in October, and the set was decorated with pumpkins that had lit candles in them. Bob pointed in the general direction of the bathroom or perhaps the director, who would take care of the situation. The little boy thought he was pointing at one of the pumpkins, and proceeded to take care of his needs right there, in front of God and everyone.

The second episode that will always stand out in my memory was when Clarabelle and the Indian had long horns and were blowing them at each other as in a duel. Zippy became very excited and started to do a typical chimp "whoo, whoo, whoo"! The chimp rocked back and forth, hair standing up all over, arms stretched out and down. When he reached the height of his demonstration, Zippy leaped toward the two characters and pushed them apart. These were his friends. He had spent hundreds of hours in their presence playing and laughing with them. He did not want them hurting each other. He played the hero and separated them the best he could and then went back to his chair.

There was a section of the Jackie Gleason show, it was called, "The American Scene Magazine." It employed short little comedy excerpts. We received a call for me to train Zippy to leap up and sit on a huge pile of dough, which was to billow up and surround the chimp with flour. He was to do it several times. The punch line of the magazine bit was, "This bread is **_untouched by human hands_**." Well, I bought several boxes of biscuit mix and several pounds of flour and set about training in my kitchen. What a mess! There was flour and dough in every corner and crevice of the kitchen and Zippy's clothes were caked with dough, but we accomplished the feat in no time.

We went to the theater in New York, and as usual, there was no time for us to rehearse.

Live on the show, Zippy was sent to the area where he was supposed to work. He screamed and "whoo-whooed" at the dough set before him! His hair stood up and he went to a corner of the area, continuing to express his displeasure. There was nothing I could do but go out and retrieve the chimp. When I got up to the area, I realized what was causing the problem. They had used yeast dough instead of biscuit dough. Yeast dough, especially in a large quantity, has a very nasty, sour smell. Zippy was repelled by the smell and made it very well known in front of a live audience. Fortunately, by this time shows were taped for broadcast later. The stage was cleaned up and <u>biscuit dough</u> was prepared. When the *live* show was over, they filmed Zippy jumping and bouncing on the biscuit dough, (which he did perfectly). The stunt was inserted into the taped version the television audience viewed, but the live audience never had the good fortune of seeing Zippy do it right.

A similar instance happened on the Ed Sullivan Show. The "twist" had become the rage and I taught it to Zippy. He was absolutely adorable doing that dance and he was perfect at it. His long arms made it hysterical. In those days, like now, it was customary to warm up the audience, prior to the actual show. Somebody would go out before the live audience and tell jokes or juggle or do whatever their forte was, to get the audience hyped up and prepared to laugh and applaud. Zippy was to open the show, so we were ready and waiting in the wings. Suddenly, Ed Sullivan asked me to bring Zippy out and have him do the twist, before the show started. The audience absolutely ate it up. They were whooping and yelling, clapping and laughing. My heart sank, because I knew what was going to happen when the actual show started. The opening intros and commercials were done, the curtain went up, twist music began and Ed Sullivan introduced Zippy. He went out onto the stage and started to twist, but the surprise was over for the audience. My suspicions had been correct. They had already seen it. They were kind and applauded appropriately, but the hysterical laughing and yelling were lost in the warm-up introduction and didn't make it into the show. It was a cute bit, but it was not the sensation it could have been. As if it weren't already a bad enough evening, my coat was stolen that same night.

For the few minutes of glory spent in the public eye, no one knows the thousands of hours spent training, cleaning, feeding, nursing, packing, traveling, car breakdowns, freezing weather, personal illness and dirty diapers, etc. that go into this business. Most of the chimps we worked with were females. They mature and start having their menstrual cycles. Their bottoms swell into a massive hump, the diapers no longer fit and wow, do they ever have PMS! They can get very irritable and cranky!

The hardest part of raising and training the Zippys, (of which there were about fourteen), was loving them too much and knowing it could not go on forever. Every day of my life, I quietly say to my animals, "Someday, you will be gone. You will have to leave me." You can't know how much I loved every animal I have ever owned. They are like my children. I have pets too and love them as well, but it is different when it is a *working* relationship and they are my "bread and butter". My life's work is invested in them. They are my creation and literally become an extension of me. The rapport that is built between me and the chimp (or more recently parrot), is uncanny. It is as though they can read my mind. They learn to read my body language, and they know the tone, not just the sound, but the intonation of my voice. If you could ever watch me work with the parrots, you would see this rapport, even more so than with the chimps. I don't even have to look at them, and they know what I want and are anxious to please me. I reach for a bird without even looking, and he steps on my hand.

No Zippy worked much past the age of eight. They grew large and difficult to carry and transport, and frankly, lost their cuteness and became a bit scary to humans. At my 117 pounds and five-foot and three inch height, it would have been easier for a 90-pound chimp to carry *me*!

The risks can increase as the chimp grows in age, size and strength. Chimps and most other animals test the authority of their handlers for any lapse. They also sense the state of your health at any given time and will take advantage of a moment of weakness. Survival of the fittest plays a part here, as well as establishing who is alpha, (dominant), in the relationship.

One early morning, around 1975, we were driving to Worchester, MA, to perform at a mall. Zippy was asleep in the back seat and I was putting on my make-up. I noticed that my skin and eyes looked yellow. Before long, I was weak and feeling sick. It turned out I had hepatitis A; the mild version, but nonetheless, I was very sick. I performed each show for three days, but had to rest in bed between shows in a nearby motel. At one point, the "Debbie Zippy" needed to go to the potty, so I got out of bed and carried the chimp to the bathroom. While I was undressing her in preparation, the chimp suddenly threatened me. Normally, I would have made a small aggressive move to put the chimp back in her place, but I was sick and had little energy, and the chimp knew. Debbie could tell I was the weaker of the two of us, and right then and there I lost my authority over her. She took control, and I knew I would never have control again, and I didn't.

After my recovery, we performed at a college, and my loss of authority was brought home even more. I knew I could never work with her again. It is an awful feeling to love, raise, train and care for an animal and then have them turn on you. She had lost respect for me, and there was nothing I could do to change it.

I never knew where the chimps went, after the terrible experience with the first, which went to the Central Park Zoo in New York City. After she was there about a year, we built up the courage to go and see her. She recognized us immediately, did all her cues for us and gobbled up the goodies we brought for her. She was overjoyed to see us, and it was a long time before we could leave. All seemed well, and we left feeling so good about it. Then, two weeks later, we received a call from the Zoo. The person calling asked, "What would you like us to do with the body?" I was shocked and asked, "What do you mean? What body?" He said, "After your visit, she wouldn't eat and grieved to death." We had to ask why the zoo hadn't contacted us earlier, but no one really ever gave us an answer. It was hard to accept that the first Zippy had died at the Central Park Zoo from grieving for us and tougher still that no one had let us know that anything was wrong. There was probably nothing we could have done, but my heart was broken and I thought I would never stop crying. After that, I preferred not to know what happened to the chimps when it came time to say good-bye to any of my little guys and put them into retirement.

Visiting retired chimps was just not a good idea for many reasons. Ralph Isaac and his wife Kathleen Quinlan, the first trainer/performer couple, went to see the Zippy they had worked. I believe the Zoo that had the chimp was somewhere in Connecticut. The chimp was tremendously excited to see him. Ralph made the mistake of getting into the cage with him. The chimp got so excited that he bit Ralph's hand severely. We learned early in the business that chimps could get their emotions confused at the height of their excitement and do unexpected things. (It could be that they confuse happiness with fear and opt for survival.)

As an example, many years ago, Curtis Candy Company sponsored what they called the "Circus Sideshow" for television. Zippy was doing the commercials as well as being featured on the show. The president of the company enjoyed having Zippy come up to his Madison Avenue office. He would have his secretary hold his calls, lock the door, and literally get down on the floor and play with Zippy. We all had a blast! The president always kept some candy bars in a small refrigerator near his desk. In no time, Zippy learned to open the refrigerator and remove candy for his own consumption. This gentleman loved Zippy and Zippy loved him VERY MUCH!

When the series was completed, we didn't see much of Zippy's good friend anymore. Months later, we took Zippy to visit him, just for fun. Zippy was so excited to see him that he whooped and whooped, stomped his feet, his hair stood up on end, and then he ran over and bit his friend. He didn't break the skin; he simply sat down and played with him again. However, this is a perfect example of how chimps get their emotions confused. Zippy felt the pleasurable excitement of seeing his friend again, but probably confused the emotion with danger and defended himself. We learned from these experiences and knew to hold Zippy back until he became calmer.

Training the Zippys was fraught with rewards and heartaches alike. One of the chimps developed cataracts. Doctors said our only hope was glasses. How do you keep glasses on a chimp? One evening, Zippy's daddy was rehearsing the chimp in the training room, while I was dressing and preparing for a show in New York City. Believe it or not, it was for the Humane Society's annual dinner at the Waldorf Astoria. I was called to the training room, because the chimp would not jump through the hula-hoop on his skates, (our final and most exciting part of the act). I held the hoop and as the chimp skated around the perimeter of the room, I said, "Jump." The chimp jumped and passed through perfectly. I tried it again, and again he passed through with ease. Suddenly, I remembered his eye problem. This time I did not say, "Jump," and the chimp skated right into the hula-hoop. He had been reacting to the sound of my voice to do the stunt and could not see! We performed the whole show for the Humane Society that night without a hitch. Then we performed the next morning at the Fowler Dick and Walker Department Store in Binghamton, NY. That was the last show we did with him. A park in northern New York took him and used him as a performer for several more years. Even though he was totally blind, he still loved to perform.

# The Zippy Act

When I first came to work with Zippy, the act was about seven minutes long. The chimp had been taught to; skate forward, skate in and out between plastic glasses, shoot his water gun, play his toy piano, ride his two-wheeler and jump into a high chair. Then he put on a bib, ate a bowl of applesauce with a spoon and drank out of a small plastic glass. That was the gist of the show.

I had no idea that I would be a good animal trainer. Other than my own dog, I had never an opportunity to test my skills. However, there is an authority and lack of fear of animals that a person either has or doesn't have. (As an example, twin men can each walk a large dog. One man may walk his dog and the other may be *walked by his dog*). I apparently had that special gift, for which I thank the Lord.

I find training an animal absolutely fascinating! The ability to have an animal do what I want it to do, and enjoy doing it, is my love. I never want an animal to perform something he doesn't want to do. No force is necessary if it is done in love and with a little animal psychology.

The rapport I have with my animals delights me to the core, and I'm told it shows in my performance and in their devotion to me.

In Flint, MI, we did a big sports show. The stage was enormous, with a huge tank of water in front of it. There was no opportunity for intimacy between a small chimp and the thousands of people that were hundreds of feet away. I had a great idea to have the chimp get close to the audience by climbing from the stage to the balcony, on a long stretched-out ladder. Then, with his water gun in hand, he was to make his way along the railing, shooting the people during the process. Upon my signal, he came back to the stage to the sound of thunderous applause. For a ten-day stretch, 2 and 3 shows a day, he did it perfectly and never even dropped his water gun.

I also delight in finding unusual props and teaching things that do not come naturally to an animal. I have spent many hours in toy stores finding wild and unusual items. I found a vehicle that is propelled by pumping with the feet. The chimps loved to perform with it and could do it expertly, backwards and forwards. Having Zippy perform with maracas, while wearing a lei and a hula skirt and dancing to the Hawaiian War Chant, was a bunch of fun! This particular twist on a theme never failed to bring the house down. I also taught the Zippys to skate backwards, spin and jump through hula hoops. They even learned to jump rope with heavy professional skates on their feet.

So many television shows and commercials involving multiple behaviors were always a wonderful challenge to me. The greatest of these were; the Xerox copier commercial, which won the coveted Cleo award, the Chiquita Banana commercial, Dr. Land and the Polaroid commercial, and last but not least, the Spiedel Watch Band commercial. There were many, many more.

Some made it, and some did not. I still remember the filming of the Mrs. Smith's Pies commercial. Zippy was to tuck his napkin into his shirt, pick up his fork and eat his apple pie. Several takes later, he was full of pie and didn't want any more. I don't think that commercial made it.

During a potato chip commercial, the person representing Tarzan could not remember his lines. All day we did take after take. That one never aired to my knowledge, but it wasn't Zippy's fault. Usually, when food was involved, the commercial was canned. (Even now, human food and animals are not a very good mix. People seem turned off or offended by it). Commercials were fun and always challenging. I usually had closer control of the chimp and retakes took the tension away.

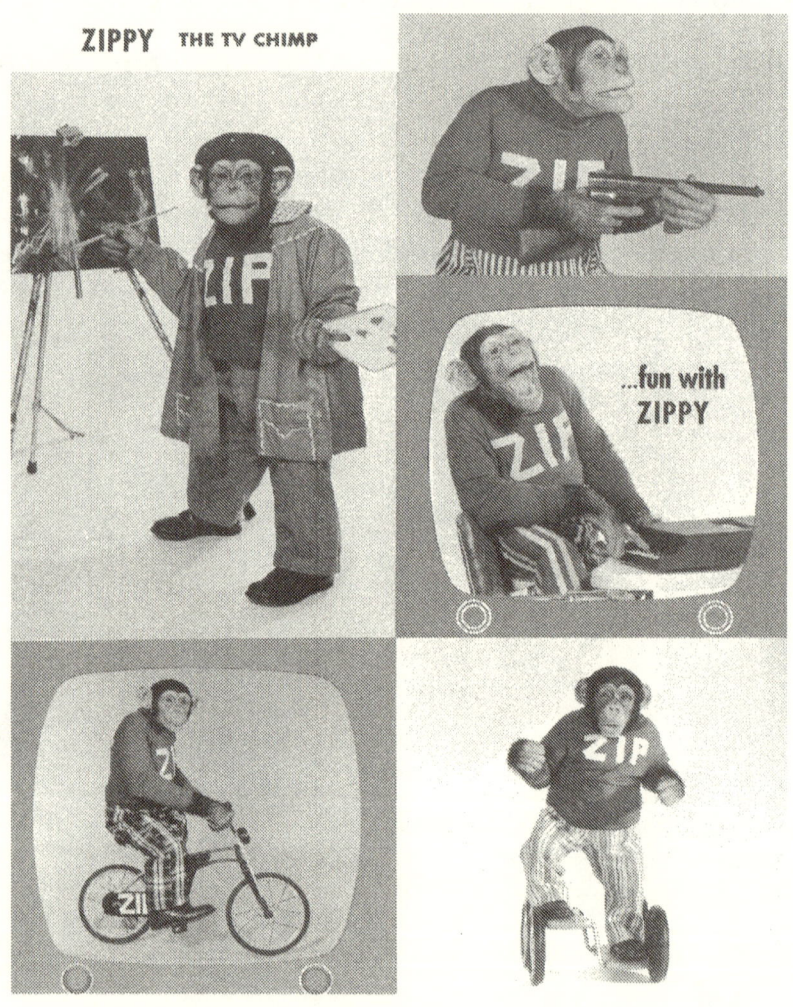

Carole's Zippy brochure (Page 1)

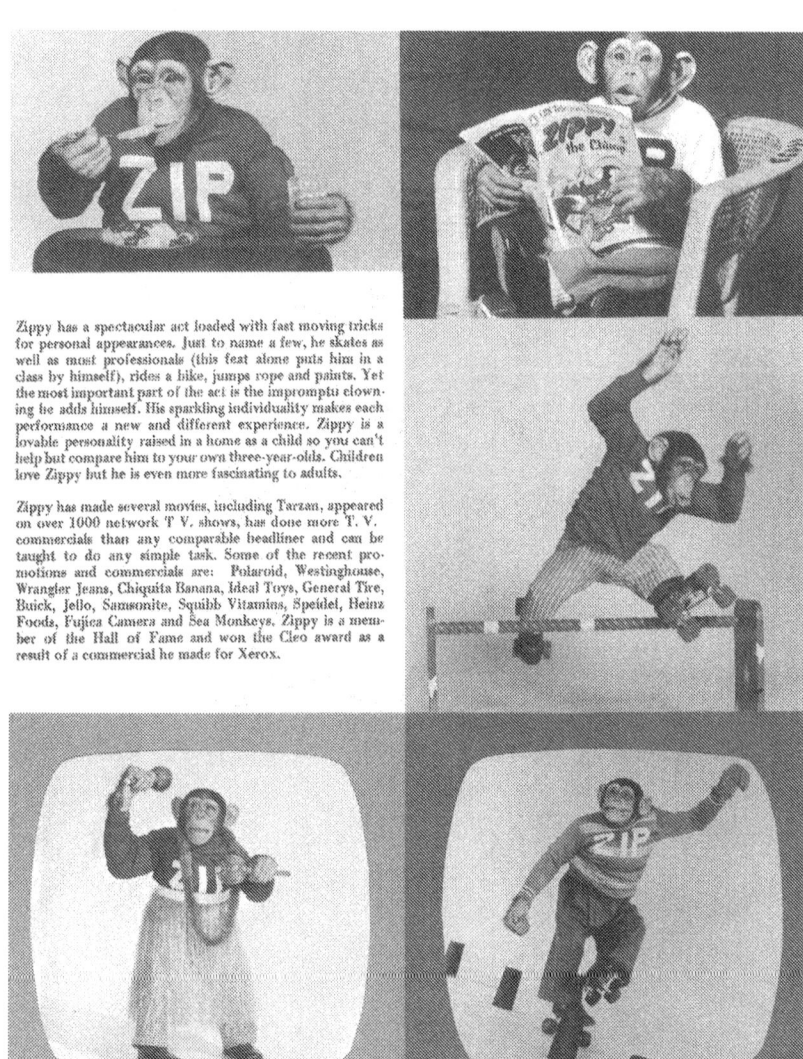

Zippy has a spectacular act loaded with fast moving tricks for personal appearances. Just to name a few, he skates as well as most professionals (this feat alone puts him in a class by himself), rides a bike, jumps rope and paints. Yet the most important part of the act is the impromptu clowning he adds himself. His sparkling individuality makes each performance a new and different experience. Zippy is a lovable personality raised in a home as a child so you can't help but compare him to your own three-year-olds. Children love Zippy but he is even more fascinating to adults.

Zippy has made several movies, including Tarzan, appeared on over 1000 network T.V. shows, has done more T.V. commercials than any comparable headliner and can be taught to do any simple task. Some of the recent promotions and commercials are: Polaroid, Westinghouse, Wrangler Jeans, Chiquita Banana, Ideal Toys, General Tire, Buick, Jello, Samsonite, Squibb Vitamins, Speidel, Heinz Foods, Fujica Camera and Sea Monkeys. Zippy is a member of the Hall of Fame and won the Cleo award as a result of a commercial he made for Xerox.

Carole's Zippy brochure (Page 2)

# Speidel Watchband Commercial

One of my greatest accomplishments ever was training Zippy for the Speidel Watchband Commercial on live TV. The phone rang and Zippy's dad, (Lee), answered. Ever curious I tried to listen and my pulse quickened when I heard, "I don't think Zippy can be taught to do that.

You see, chimps have a non-opposing and underdeveloped thumb, (compared to humans), and for this reason they do not have as much dexterity as a human would have. I really don't think we can teach him that." I just couldn't resist interrupting and asking what this was all about. Lee explained to me that it was the Speidel Watchband Company. They wanted Zippy to tie a knot in a Speidel Twist-o-Flex watchband. Lee didn't think Zippy could do it and had decided to say no, but I said "Well, why don't you let me give it a try? Nothing ventured, nothing gained." So Lee returned to the phone and said, "My wife says she wants to give it a try." The next thing I knew, several men were in our home with a variety of watchbands. Naturally, everyone wanted to see the chimp, so they had brought their wives along to help them deliver the bands.

Training was a long drawn out process, and I literally ran into a "catch." The bands had a spring-loaded pin attached to each end, which tangled with every attempt at knotting. It was quite a challenge, and both Zippy and I found it to be very frustrating. So we filed down the ends of the pins, ever so slightly, and began to make some progress with Zippy's training. I worked day and night teaching him this task. While riding in the car, before meals, after meals, before bedtime, we practiced and practiced.

Finally, we were ready! We arrived at the advertising agency with Zippy to prove that he had learned to tie the band into a knot. Of course, everyone in the agency wanted to see Zippy and his accomplishment. Zippy sat in this big stuffed chair and successfully tied the knot at least a dozen times without failing even once. People were flabbergasted to see how nimble and talented he was.

I was asked what I would need as props. After hearing a description of the basic set up, I asked that they provide a stool with a back on it. Zippy was to leave his seat in the audience, ascend the steps to the stage and climb onto the stool. Then Jackie Gleason would come to his side, hand him the watchband and Zippy would tie it into a knot. I suggested a stool with a back on it to ensure that Zippy would sit facing the audience.

The date was set for the first commercial to be done live on the Jackie Gleason Show. As usual, rehearsal was running late, and Zippy was saved until last. Finally, it was time for Zippy to take the stage. The director told me to have Zippy in the audience and when it was time, to send him up to the stool where Mr. Gleason would meet him. I saw them bring the stool out, but much to my dismay, it had no back! By then it was too late to get a different stool. Then the FCC agent came to me and asked to inspect the watchband that Zippy was to use. I produced the band, and he said, "This watchband has been tampered with." I explained that we had to file the ends down a little so that they would not catch as Zippy tied the knot. He then told me that we would have to use a new band in order to comply with FCC regulations. I started to worry that we were headed for disaster that evening, but we had no choice other than to comply.

Zippy and I sat in the audience anxiously waiting, and I was dreading what might happen next. I felt like a mother whose child was about to give his first recital. When called, Zippy left me, went onto the stage, and climbed onto the stool, facing the crowd. Well, we had made it past the first hurdle. Mr. Gleason's stand-in walked over, stood next to Zippy and handed him the watchband. Zippy proceeded to tie the knot without a hitch! No mother has ever been more proud! Then, unbelievably, the director asked me to have Zippy hold the band up high, so they could get a better shot of it. I was astounded! Did he really think the chimp was human, and all I had to do was ask for the chimp to do it? I had no choice but to try. So I complied and showed Zippy what they wanted, as though I really expected him to understand and cooperate. The first time, Zippy misunderstood and threw the band. The second time, he handed it to the stand-in. However, on the third try, to my astonishment and pleasure, he did it!

That night Zippy and I sat in the audience with people all around us. On cue, little Zippy went onto this enormous stage, got onto the stool and faced the audience. The next surprise for Zippy had the potential to be the worst. Since Jackie Gleason never attended rehearsals, Zippy had not worked with Mr. Gleason, only his stand-in. Animals tend to learn tasks very specifically, and any change to a routine can cause confusion. Zippy was being asked to perform an extremely difficult behavior with a "stranger" for the actual live show. Would an animal, even one as talented as Zippy the Chimp, be able to overcome these obstacles? Jackie Gleason walked over and handed Zippy the watchband. The chimp tied it and held it up just as he was supposed to do.

When the spot was finished, Zip returned to his seat next to me in the audience. Perfection! What a moment of accomplishment and pride for Zippy's mom.

Zippy did this same commercial live six times on "The Price Is Right", and he did it perfectly every time. Bill Cullen, the host, really enjoyed working with Zippy and being his friend. Jackie Gleason also used us several times after that in many different skits.

In one of his skits, Jackie sat in a chair next to Zippy. A cup of coffee was to be served to Mr. Gleason by one of the glamour girls on the show and it was decided that I was to be that girl. I was supposed to be dressed in a French maid's outfit. I went to several costume rental shops, but could find nothing tiny enough. So I created a very attractive outfit myself. In the show, I brought out the coffee and placed it before Jackie Gleason. Zippy looked at me as if to say, "Hey Mom, where's mine?" I returned to the wings, Zippy finished the skit and walked over to me, right on cue. What a chimp!

After the Jackie Gleason Show was over, Mr. Gleason had a 13-week show in which he interviewed people. Two big overstuffed chairs were separated by a table, in much the same way as you might arrange your living room. Jackie decided he wanted to include Zippy. We were called to his office to do some brainstorming. I took Zippy to a set of chairs in the corner of the office and gave him an apple to eat in order to keep him occupied while the men were talking.

Chimps do not eat the skin of apples. So each time he finished a bite, Zippy would spit the skin into the ashtray I was holding for that purpose. We developed a rhythm to this action. When I thrust the ashtray close to his face, Zippy would spit the skin into the ashtray. If I became distracted and did not do my part at the expected time, Zippy would reach out, take my wrist and bring my hand (and the ashtray), close enough to spit the skin into it.

Jackie Gleason was watching all this from his desk, and suddenly he pointed to us and said, "We'll do that on the show!" Everyone was puzzled and asked in unison, "Do what?" Then he said, "We'll have Zippy eat an apple while I interview him."

The night of the show arrived, and as was his custom, Jackie had an understudy practice with Zippy. When the show began, Zippy went to his chair. Jackie introduced Zippy and handed him a huge apple, (it was going to be a long bit). Jackie did a great narration. Zippy made faces, did the raspberry on cue, nodded in response to Jackie's questions and ate his apple. Jackie dutifully offered the ashtray, (becoming overloaded with skins), to Zippy as agreed. Near the end of the interview, Zippy had only the core of the huge apple remaining. Jackie was distracted and forgot to hand the ashtray to Zippy. Zippy was unable to reach Jackie's arm and became frustrated. So he tossed the core into the air. Amazingly, it landed in the ashtray, scattering chewed skins all over the place. Everyone on stage and in the audience went wild and the bit was a sensation.

Zippy was always a big hit with the audience and with Mr. Gleason as well. Once Jackie even invited Zippy to his favorite haunt, "Toots Shore's", where Zippy joined him for a drink, (a soda for Zippy of course).

# The Box

Joe and Debbie Buglisi worked with Zippy for several years. They were beautiful people and good friends whom we loved dearly. They appeared frequently on the Merv Griffin Show. I particularly remember Milton Berle and Zippy having a painting contest on Merv's show. It was an adorable bit. Wherever Debbie and Joe went with Zippy, they were well received. They handled their chimp expertly.

One of the most innovative things ever done for the chimps was simply a box. Joe's father designed it. It was a box made of solid wood and was approximately 2 ½ feet by 3 feet and about 2 ½ feet deep. The box consisted of 6 pieces which were dove tailed all the way around each piece so that each piece meshed with another. There were notches large enough to hold 2-inch wide webbed nylon straps in place to anchor it all together. One went around the girth and 2 straps around the top and bottom. The box was riddled with air holes.

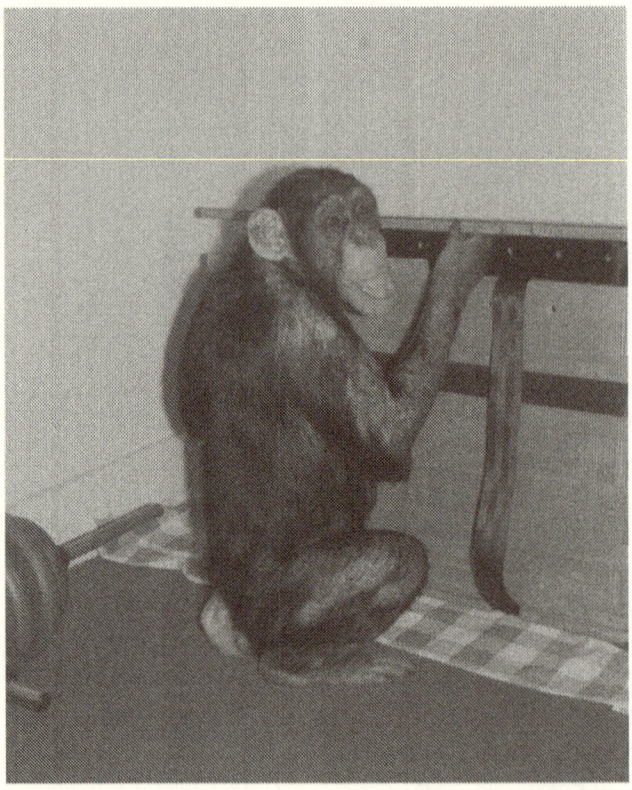

Zippy and "The Box", his home away from home

The box went everywhere with us and the chimps loved it. While staying in hotel rooms I would assemble the box, bathe the chimp and put his blankets and such in the box. The chimp would jump right in, make a nest, make his goodnight noises with grunts and coos and get a kiss goodnight. The lid would be strapped in place and he was out until morning. He would knock softly to let us know when it was time for him to get out and go. We always hurriedly un-strapped the box, gave him a big hug and took him swiftly to the toilet. He usually had to go immediately. In all the years we used the box, we never had an accident in it.

This box greatly simplified life for us. We were free to go out to dinner or a movie without worry. Not only did the box afford us some freedom, it was a place of security where the chimps could sleep without fear. Each was eager and relaxed entering the box at night and slept comfortably and restfully.

There were times we utilized the box on air flights. We insisted on watching the box as it was put in the cargo area of the plane. The box was to be the last item so it could not be covered with baggage and suffocate or crush the chimp. One night we watched in horror as Eastern Airlines closed the cargo door on the box and crushed the corner. Fortunately, Eastern repaired the box, but we traveled by car after that.

People were surprised after the box arrived at baggage pickup. We quickly pulled the box off of the conveyor, un-strapped the box and out would pop a naked chimp. The chimp and I rushed to the nearest ladies room with a bag of clothes and his roller skates slung over my shoulder. Zippy was good about holding it till we hit the potty. In 5 minutes he was dressed and skating through the door of the restroom. That was a show to behold and a great source of pride in our Zippys.

# The Xerox Chimp

We called him the Xerox Chimp because his commercial won the Cleo Award. This is the highest honor available for television commercials, and I trained this chimp. Yes, I am proud of that achievement.

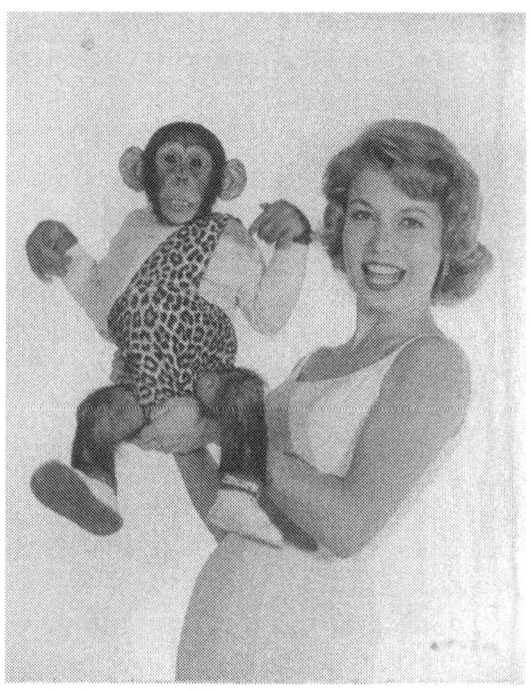

Publicity shot for a Tarzan movie

After being booked by the advertising agency, a portion of the training commenced. Other than training the chimp to swing on a rope and hand-off and receive papers, there was little I could do without the actual copy machine, which was about the size of a dining room table. There was no opportunity to bring it to a studio or other such place accessible to me because the cost would have been prohibitive. Therefore, most of the training had to be done the same day as the actual shooting.

The commercial opens with a business man sitting at his desk. As Zippy swung across the desk on a rope to the other side of the screen, the business man handed Zippy some papers. The man said, "Take these and get them copied," without looking up at the chimp. The next shots showed Zippy running and holding the papers, hopping up onto a row of chairs and running across them. Then he ran through and over several more obstacles. Finally, he reached the huge machine, jumped up onto it, opened the flap, inserted the paper properly, closed the flap, reached over and pushed the buttons <u>in proper sequence</u>, jumped down, and went to the other side of the machine where he retrieved the papers from the collator. Then, he returned with the original and the copy in one hand, in and out of the office type obstacles, and again swung over the businessman's desk, and handed the man (who didn't look at Zippy), the papers and swung out of the scene. All this was accomplished in one swing. The business man looked at the papers and said, "Which one is the original?" All this he did in one take. We only rehearsed this segment for about an hour during the lunch break.

This commercial was shown on television during a massive production several times that night. The next day, Xerox began to receive phone calls and complaints from secretaries all over the United States. These complaints came from women who were finding bananas on their copiers with notes saying, "If Zippy can work the Xerox machine, why can't you?" To my knowledge, the commercial was never run again, but it still won the coveted Cleo Award.

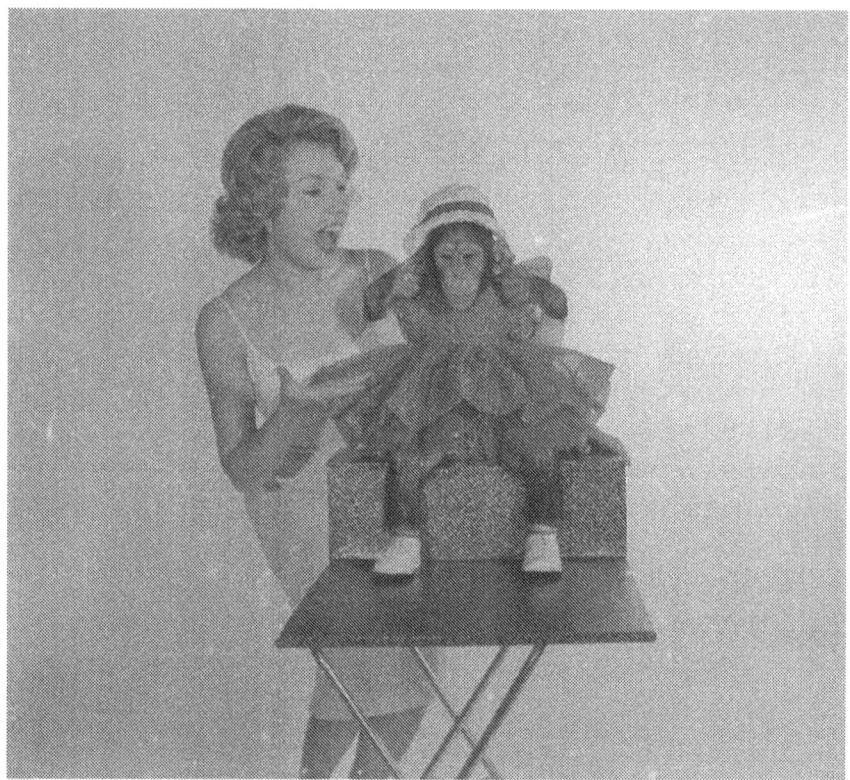

I've Grown Accustomed to your Face

In the Chiquita Banana commercial, this Zippy pushed a huge bunch of bananas through a grocery store. This is the same chimp that did the twist on the Ed Sullivan Show. He also did a Lays Potato Chip commercial, The Jackie Gleason Show, (he jumped up and down on the dough in that magazine segment), and many other commercials and shows too numerous to recall. He was also the chimp that toured with Ian Sutherland around New York (another chapter).

This little guy was vibrant! He put all his energy into everything he did. I bought maracas by the dozens, because they could not withstand his enthusiasm. Zippy donned his hula skirt and lei and I handed him his maracas, while the orchestra played the Hawaiian War Chant. Zippy danced and spun on his heel unbelievably fast. He shook the maracas so hard that they would often hit each other and break. We glued them and taped them until they could be repaired no more. Hula or Twist, no chimp ever danced like the "Xerox" chimp.

One of this chimp's favorite toys was a front tire from a private plane. It was looped on a chain that hung from the ceiling of the play room. All the chimps enjoyed twisting the tire as if winding a spring and then riding it as it unwound. One day I went in to serve lunch and saw this chimp hanging there with his arms wide spread. At first I thought he was going to jump into my arms, but suddenly a chill went over me. He had accidentally wrapped the chain around his neck and was dead. We tried for hours to revive him, knowing in our hearts our efforts were futile. We wept until exhaustion overcame us. We laid him on our bed and returned later to find that by some miracle he had revived, but to no avail. Of course, waves of guilt flooded over us, and the tragedy never left us. To this day I am shaken when I remember that horrible time.

# Polly Coffee and Mrs. DuPont

One day we received a call out of the blue from a wonderfully zany lady named Polly Coffee. She called to ask us to perform for her Christmas Party in New Jersey. Apparently she had seen Zippy on television and thought he would add an extraordinary touch and really liven up her party. We sent her a contract and soon, one holiday evening, we found our way from Long Island to her home (mansion is a more accurate description) in New Jersey. It was even more grand than expected.

We arrived at the back door, as instructed, and were escorted up to one of the huge bedrooms, which we were to use as a dressing room. Before long, Polly Coffee came bursting through the doorway! She immediately grabbed Zippy, picked him up, and hugged him close to her while spinning around and around. Needless to say, Zippy fell in love with her on the spot! Chimps seem to sense immediately the authority, love and sincerity of people. There was nothing at all phony about this woman. She loved animals and they knew it! Polly did everything in a big, grand manner and she did it *her* way. What a great lady!

After she left us, Zippy and I prepared ourselves for the show, while Zippy's dad set up the props and sound system. Soon he came up to join us and changed into his tux. Finally, it was show time!

We walked down the long hall and proceeded down one of the two huge, sweeping staircases which surrounded the enormous, marble entrance hall. There we found more than one hundred formally dressed guests waiting for the entertainment to begin. They were sitting on both stairways and some were also on the balcony looking down. Our "stage area" lay between the stairs.

The intro music started and the show began. Zippy skated in and whizzed around that marble area. He danced, rode his two-wheeler, painted, jumped rope, plus performed all his other wonderful bits. But when it came to his water rifle, I was more than a little nervous. All those gorgeous, formal evening gowns were going to get wet! He voraciously and enthusiastically squirted water until he had no more! I cringed as many of them were drenched, but no one complained. In fact, they loved it! I would suppose that friends of Mrs. Coffee should expect such fun at one of her parties. Suddenly Zippy spotted Mrs. Coffee. He sped over to her with that gun and rapidly squeezed! Polly loved it. She ran around the entrance hall stage area and played a kind of tag with Zippy, both of them dodging and laughing all the while. If it had been planned, it would not have been more precious.

I was later told that Mrs. Coffee was, at that time, the ninth wealthiest woman in the world, with most of her fortune coming from the lumber industry. For all her great wealth, there was nothing the least bit stuffy or pretentious about her. Polly hired Zippy for parties on numerous other occasions after that. At Christmas time, she had us hire other acts for her as well. She always preferred animal acts.

Once, we hired an act using trained penguins and seals, but the Christmas act I will never forget was a cowboy with his "Wonder Horse". His horse was trained, and the cowboy did guns and whip stunts as well. It was very cold that Christmas, and there was snow on the ground. The show opened with the horse act. The huge double doors at the entrance to the hall swung open to the back of the stage area, and in rode the cowboy on his horse. What a grand entrance indeed. The horse reared up, as he normally did for his opening, but then he suddenly let go. (What he let go was a huge steaming pile)! The crowd got very quiet for a moment, but then everyone began to laugh as Polly nonchalantly summoned one of the maids to scoop it up. The mess was discreetly scooped up and the show went on as scheduled. Mrs. Coffee never so much as blinked an eyelash.

Then it was time for Zippy to do his show. Yes, you've got the right picture. Zippy immediately went over to the area of the horse's movement and sniffed it. His hair stood on end and he did the traditional "whoo, whoo, whoo", that chimps do when they are upset or bluffing. As described earlier, it starts low and increases in multi-decibels, getting louder and louder, and ending in an ear piercing scream. His verbal action is followed by stomping his feet and doing the typical Tarzan arm gestures. It seemed like Zippy was asking who had done this terrible misdeed, as if he were insulted and disgusted by the whole thing. The audience screamed with delight at Zippy's quite natural improvisation. Then the maid returned with a pail and mop, and soon all was back to normal (or at least as normal as it gets).

Mrs. Coffee also had two homes on Fischer Island, which is located in the Long Island Sound. There are only two ways to get there; via boat or small plane. Mrs. Coffee requested our presence there on one particular occasion. She offered to pick us up at a small airport in the middle of Long Island. We accepted her offer, so one summer evening we met at the airport and loaded Zippy's props, costumes and ourselves into her plane. In addition to all her other amazing qualities, Mrs. Coffee was actually a licensed pilot. She and her co-pilot were sitting in the front two seats, Zippy, his daddy and I squeezed into a double seat; and three of her children sat across the back.

Once we were all loaded and reasonably comfortable, we took off. Zippy was enjoying the ride, when suddenly I looked over the co-pilot's shoulder and saw that he was looking at a Long

Island map, trying to identify the Long Island Expressway. I realized that he was using this map to navigate, which seemed rather strange and not particularly comforting to me. The sun was quickly setting as we turned north toward the Island. I looked out the window, saw no lights and felt we were very close to the water. Suddenly the stall alarm went off, which is not a sound you want to hear when you're in a plane. Polly simply reached over and slapped the dashboard while shouting, "Oh, shut up!" With that, she gunned the engine to lift the nose of the plane and barely cleared a levee of dirt or concrete placed there to keep high tides from coming onto the landing strip. We cleared the hump and sort of "belly flopped" onto the runway without further movement.

A van came down the runway, loaded us and our baggage and delivered us to her mansion. Our "job" was to simply mingle with her guests during the cocktail hour. People enjoyed handing Zippy their drinks and watching as he took a sip from each of their glasses. Many of these were alcoholic drinks, so I was careful to make sure he only took a sip here and a sip there. An intoxicated chimpanzee would be asking for trouble (plus, I would never do anything that might be harmful to him in any way). Later that evening, we were taken to another of her homes on Fisher Island, but this house was basically for her children. They were having their own party, and we performed our show for them. The children loved Zippy, and he enjoyed every moment as well.

After that, we returned to the first home, but we could not find Mrs. Coffee. We were supposed to return to Long Island, but by the time we spotted her it was after midnight. She gave us a choice: she could fly us back or we could stay the night. Considering the time and how the ride over had gone, we opted to stay. We were driven to a small guesthouse that belonged to one of Mrs. Coffee's neighbors. We had not planned on staying the night and had brought none of the necessities. Two single beds were in the room. There was no box or bed for Zippy, so he climbed into bed with me. I thought it was a good idea at the time; at least I would know where Zippy was and he wouldn't be out wandering around getting into trouble. BIG mistake! You see, I'd forgotten about all those drinks he had sipped earlier that evening. Much to my surprise early the next morning, I awoke in a puddle.

We bathed and got dressed in our clothes from the night before. We opened the door and discovered that our quarters were located near the waters' of the sound. A long dock extended from shore, with a schooner on one side and a motor-yacht on the other. To the rear of the guesthouse stretched a perfectly manicured lawn for about as far as the eye could see.

Having no other options, we walked up to the mansion on top of the hill.

As we drew nearer the house, we could see activity on the screened-in porch. At closer range, we could see a huge table with about 12 people seated around it. We were met at the screen door by several children, who gleefully escorted us onto the porch. The lady seated at the end of the long table, Mrs. DuPont, invited us to join everyone for breakfast. I can't honestly remember her first name, but one doesn't forget the name DuPont. She was a lovely, youthful lady; graceful and slim with long blonde hair. Apparently, it was in her guesthouse we had spent the night. What an embarrassment it was to explain to her what had happened to the mattress, but she was very gracious about it. We enjoyed our breakfast, posed for many pictures with all the children, and then Mrs. DuPont took us to the plane. During the ride we shared some great conversations, but as all good things do, it came to an end. Another pilot flew us back to the airport, and we made it home safe and sound. Entertaining at one of Mrs. Coffee's parties was always such an adventure.

# Zippy's Friend Ian

Our next adventure included a special friend of ours, whose name is Ian. Even though Ian was a stockbroker in New York City by trade, he also knew how to relax and have a good time. He and his wife, Erika, would sometimes join us on our boat. We would water ski or fish, and play with the chimps.

On one of our outings together, Ian asked if we could all go into the city. He wanted to see for himself if all the stories about our many escapades with Zippy were really true. So we all dressed to the hilt, strapped on Zippy's skates and went to Sardi's, the famous restaurant of the stars, we had eaten with Zippy there on many occasions. You see, Ian thought watching people re-act to Zippy's wonderful table manners was the delight of his life. Of course, heads turned from the moment we entered the restaurant. Zippy's manners were impeccable as always, and everyone was amazed to see that he could actually use eating utensils properly. It wasn't long before people began coming over to our table to introduce themselves and meet Zippy in person. He was Sardi's star attraction for the evening.

We all had a delightful time, but that wasn't enough for Ian. There was a Broadway theatre nearby where Sir John Gielgud was starring in a play. So off we went to see if we could get tickets. The manger of the theatre granted permission for us to purchase tickets in the prime spot of the theatre; fourth row center. As we entered the theatre, the lights were on and the theatre was full. Zippy skated down the aisle, stopped on cue and waited for us at the fourth row. I was accustomed to people staring at Zippy, but was surprised when I heard a great rumbling sound from above. I turned and saw that all the people in the balcony above and behind us were pushing their way up to the rail to get a better look at Zippy. Some were even leaning over the balcony as they strained to see what was happening. I became frightened as I realized that the rumbling sound was coming from the balcony itself, as if it were trembling and creaking from the weight of all the people pressing forward. I got a sinking feeling, as I feared it might crumble from all that weight, or that someone would fall over the rail of the balcony.

Zippy, however, wasn't phased in the least. He was accustomed to crowds and theatres and being the center of attention. We quickly took our seats and I gave Zippy my fur jacket and a couple of his spare shirts, which we kept in a briefcase for him. As was his custom, Zippy promptly began making a nest in the seat and quietly curled up for a nap.

Seven o'clock came and went. The lights were low, Zippy was asleep, and we were all waiting for the performance to begin. Nothing happened. We waited and waited until the audience became impatient and began fidgeting and murmuring. Then suddenly we realized that the manager of the theatre was in the aisle trying to get our attention. As quietly as he could, he passed on a message from Sir John Gielgud. It seemed that he didn't want to perform for a "monkey", and the show would not begin until we left. Naturally, I was offended and hurt by the unfairness of it all. Zippy was doing nothing wrong. In fact, he was behaving better than many children would have been. Everyone in the audience had been charmed by him. Perhaps Sir John was afraid that Zippy would upstage him and make a "monkey" out of him instead.

Not wishing to create a scene, we gathered all our things, woke Zippy and made our way toward the aisle. As we started toward the rear of the theatre, the audience began to notice us again. I guess a chimp in roller skates is pretty hard to miss. They began stomping their feet and jeering when they realized we were being ejected. It seems they didn't want Zippy to leave. The "boos" and stomping grew louder, and the rumbling sound started again. I had a very unsettling feeling again when we reached the area under the balcony, which I prayed would hold up under the weight of the stomping crowd. We hurried toward the exit. Our money was refunded and we left the theatre.

From there we went on up the street to the NBC building, where Zippy had his own dressing room. He knew the layout of the building well and led the way. We decided we would try to get him onto one of the shows being filmed. Zippy, always the ham, wound up helping an astonished anchorman do the weather. He would even point to the numbers on the map. At the end, the anchorman leaned the microphone down toward him and asked him for his own personal forecast for the next day's weather. Zippy gave a loud raspberry into the microphone. As usual, Zippy somehow managed to steal the show. We had an eventful evening, but for us it was just another night out on the town in New York City. It was the night we made Ian a believer.

# Training Procedures

Positive reinforcement is the key to great training. It not only discards the need for punishment, but also brings out the "ham" qualities of the animal. They quickly pick up on what makes an audience laugh and what pleases or upsets me. Therefore, training periods become shorter, and behaviors become more and more intricate and taught with ease. People often say, "You must have so much patience." No, it's not just patience. I think it is more that I understand positive reinforcement and rewards. Also, I push to see how far I can go and set goals far beyond what I honestly think the animal can do. When he does it, I'm more surprised and thrilled than anyone and let my excitement and praise show to the animal. Animals, like small children, have a short attention span. Training periods should be kept to a minimum, lasting about 20 minutes at a time. Usually, four or five training periods were scheduled each day, with eating and play time allowed in between.

Sometimes it is hard to quit, especially when on the brink of accomplishment. It is also hard to quit when frustration dominates the situation. I recall one time when I was teaching a chimp to skate in and out, between plastic glasses. This was a standard part of the act. Zippy would skate through the cups like an obstacle course and then pick them up and stack them one at a time. The obstacle course portion of this dual behavior makes no sense to a chimp, so it is hard to teach. Constant repetition is the only way. On this particular occasion, I was frustrated because the chimp kept sitting down. Knowing that I, not the chimp, was out of control, I sat down on the couch in the training room and removed the chimp's shoes. These shoes were high top shoes that supported the chimp's ankles for skating. There is a metal fixture in the heel for the detachable skate to hook into that holds the skate in place. When I removed his shoe, I saw blood on the chimp's heel. Upon closer inspection, I saw that the tip of a screw, which held the skate fixture in place, had penetrated the padding and was irritating the chimp's heel. That was why he kept sitting down! I was so very grateful that I had not forced the issue and I have never forgotten the lesson; when an animal rebels, there is usually a reason.

A similar incident happened with Cocoa, one of our Blue & Gold macaws, while working a show at the State Fair of Virginia. One of my favorites, Cocoa was to roll over on the table. She didn't want to do it. I tried to rehearse with her between shows, but she still would not do it. During the next show, she started to roll over, quit and then ran over and bit me. She bit me seven times during that show. Fortunately, it was the last show of the day. My husband, Don, arrived that night. I showed him my wounds and told him what had happened. He checked Cocoa out and found a badly infected feather follicle under her wing. He cleaned it out, and later a veterinarian cleaned it again. Eventually it healed perfectly, but that bird never trusted me after that day. She thought I was the one instilling the pain whenever she rolled over onto that hurting wing. That was a horrible day for both of us! Recently, Cocoa has accepted me again and now even comes to me to be petted. We still own and care for ten of the birds, we are all retired now.

I fancied myself a great trainer and really taught some unbelievable feats. Skating, riding bikes, and even doing the twist were common behaviors. One of our trainer/performers, Fran Rinsky, taught Zippy "Chinners" sign language. This was the last chimp I trained and Jack and Fran were the last couple, as well.

After Don Womack and I had married and were already working with the parrots, Jack, Fran and Chinners came to visit with us down south. That chimp was unbelievable! He could sign for whatever food he wanted, sign to go to the bathroom, and even signed to Don when he wanted a hug. Zippy adored Don. He would sit and watch television with our three daughters, suddenly get up, turn to Don, and sign that he wanted a hug. If Don held his hands out, Zippy would run over, give Don a quick hug and then run back to his place to watch more TV. If Don did not hold out his hands, Zippy would whine and protest. Sometimes he would do this three or four times in an hour. The hugs were always very quick. Although he wanted the touch, he didn't want to remain for a long period of time. The signing really impressed me. I had seen this taught on television in research laboratories, but never seen it incorporated into an everyday life and totally volunteered by the chimp.

At dinner, Zippy ate with us all at the table. Zippy would always tuck his napkin into his turtleneck shirt, and feed himself with his fork or spoon. This was something we had taught him to do, but something new had been added! When he was finished, _without being told,_ he took his dishes over to the sink, deposited them on the drain board and threw his napkin into the trash. What a surprise! The chimp had developed better manners than some people I've known.

A chimp has a small thumb, and as a result, any task requiring a lot of dexterity is difficult for him. When Fran and Jack further trained their Zippy, (Chinners), he was buttoning his shirt and tying his own shoelaces. She took training chimps even further than I had thought possible.

# Merchandising

Merchandising started when Zippy was a regular on the Howdy Doody Show. Mrs. Mary P. Rushton, owner of the Rushton Company in Atlanta, Georgia, came to New York to see Zippy. At that time, the Rushton Company was the largest plush doll company in the world. In Zippy's dressing room at the NBC studios, Mrs. Rushton used modeling clay to create her perception of the unique face of the Zippy doll. In August of 1954, the doll was introduced, and that following February, it was presented at the Toy Show in New York City. The Zippy doll was voted "The Best Doll of the Year." The original ZIP, (Zippy the chimp), doll had a red and white hat with Howdy Doody's likeness and name on it. The trademark "ZIP" was on the front of his yellow shirt.

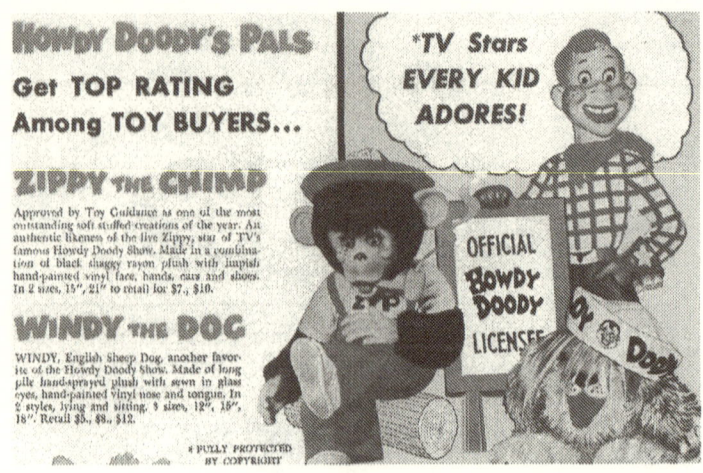

1950's original Zippy doll ad

Although this doll is now very valuable as a collector's item, the larger 20" doll retailed in 1954 for $10.00, and the smaller 15" doll for $7.00. This doll was probably the most beloved plush doll in a child's collection at that time. After the Howdy Doody Show ended, the Zippy doll continued to be marketed, but without the Howdy Doody picture and name on the hat.

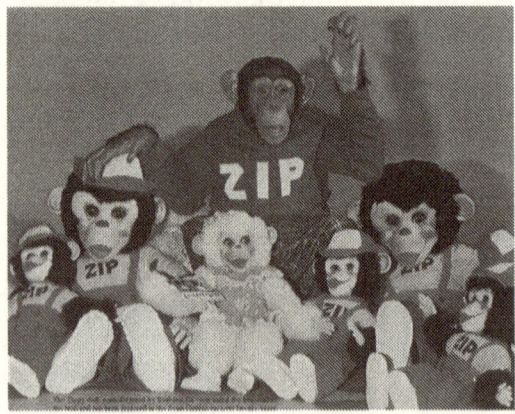

Tippy and a few of the Zippy dolls

While Zippy was a regular on the Howdy Doody Show, many merchandising items came out. I recently saw the Zippy marionette advertised on eBay. It is terribly ugly, and I confess that I would not purchase it for my children. To my astonishment, it sold for almost a thousand dollars.

A short while after the Rushton doll was released, many replicas hit the market. Mrs. Rushton did her best to stop the production of all the copies, but there were just too many. Some Asian pirates were so brazen as to actually copy "Copyright The Rushton Co.," which was marked on each original shoe. The Zippy doll remained a favorite for a very long time. It was featured in the Sears catalog for over 20 years. Mrs. Rushton made several variations of this doll. An even larger version of the doll was produced, but at that size was not quite as cuddly, even though it was proportionately the same.

A "Davy Crocket" version of the Zippy doll, complete with a toy shot gun, was one of my favorites. A cowboy doll and other dressed-up Zippy dolls were marvelously costumed. I suspect the only reason they didn't stay in the line very long was because there was no "ZIP" on the costumes. Only a small number of these Zippy dolls were ever produced, making them worth a mint in today's market. A smaller Zippy doll with "Z" on the chest and sporting a red "Superman-type" cape was the last one produced. I am fortunate to have this doll in my collection.

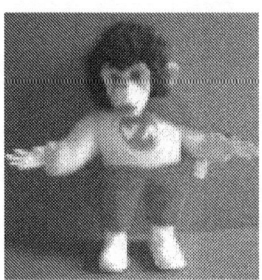

Super Hero Zip doll

Somewhere along the way, a simpler and probably more economical doll was produced. It did not have removable clothing, and the body was rather stiff, but it definitely had the Zippy face. The upper torso was yellow, and it appeared to be wearing plaid overalls and a collar with a snap on it.

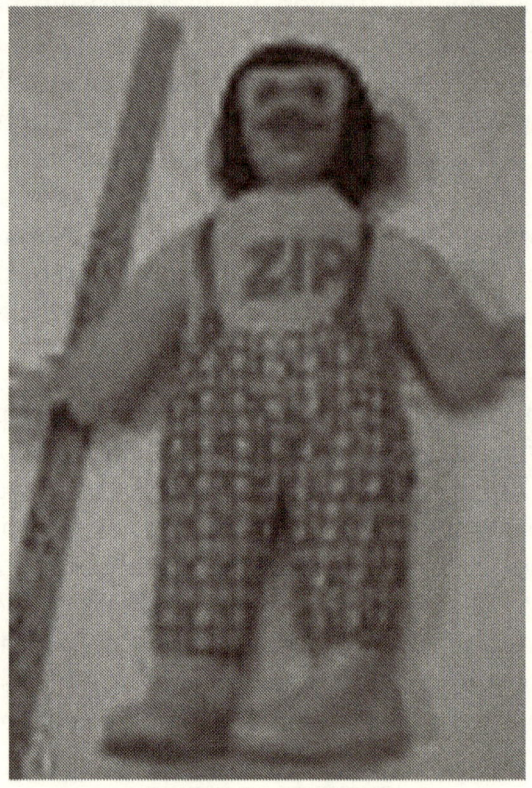

"Economy" Zip doll in checked pants

Mrs. Rushton decided that Zippy needed a girlfriend, so the Tippy doll was produced. She was blonde with a red bow in her hair and wore a red jumper with the name "Tippy" on its front. Of course, she had the same face as the Zippy doll. There was a noisemaker placed in the doll's wrist that squeaked when you pressed it. Something about the sound of that squeak seemed to fascinate children. To this day, people tell me that they recall the doll they owned and refer to the sound it made and how it always delighted them.

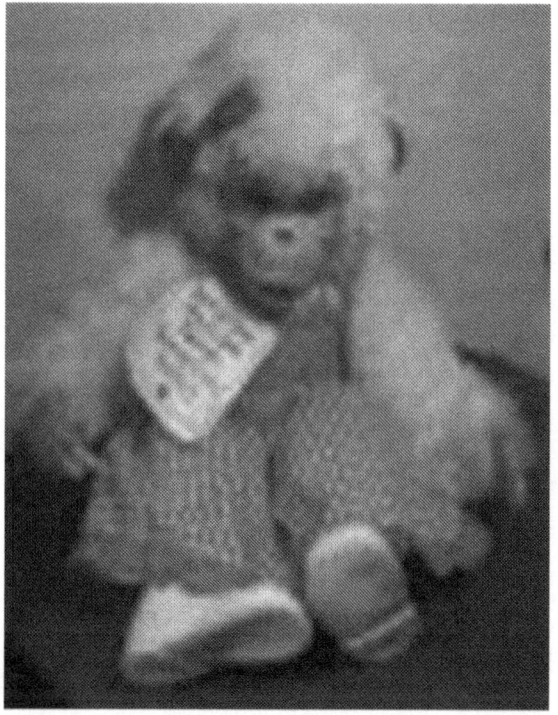

Zippy's girl friend Tippy

A few months ago, I was watching a Zippy doll being auctioned on eBay. I found the e-mail address of the winning bidder and wrote to him. I told him I was Zippy's "mother" and asked him why he paid **$162.00** for the doll. He said that he was surprised to hear from me and shared the following story.

"When my brother and I were little, we each had a chimp doll. Mine was a J. Fred Mugs doll and my brother's was a Zippy doll. These dolls were our constant companions. They ate with us at the table and went everywhere with us. Years later, our mother disposed of our old worn out dolls, and my brother and I were devastated! In a few weeks, it will be my brother's birthday. I wanted to send him a Zippy doll. I knew he would love it! I think it would be a hoot if you would send him a birthday card from Zippy's 'mother'. Would you do that for me?" I sent that card, and later I was privileged to read the outcome of the gift and the card. It was awesome!

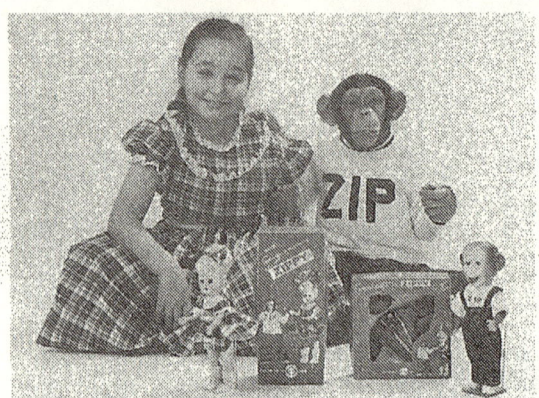

## Cosmopolitan's New Zippy Doll Set for Coast-to-Coast Tour

The live and doll versions of Ginger and Zippy the Chimp

Cosmopolitan Doll & Toy Corporation, manufacturer of Ginger dolls and dresses, will sponsor a coast-to-coast tour of Zippy, the frisky chimp seen by millions over the CBS-TV network.

Kathryn Kay, sole sales representative for Cosmopolitan, has organized the tour as part of the extensive promotion campaign for "Zippy the Chimp," an eight-inch playmate to the Ginger doll.

"Zippy is a sure-fire attraction for toy departments," says Miss Kay. "The children will be able to see that the playfulness of the live Zippy has been successfully transferred to our Zippy."

Owned by Lee Ecuyer, Zippy will put on a 15-minute show for the children at his appearances. He will skate, dance, pound the piano, play tag with the children, and give away the all-vinyl, fully-jointed "Zippy the Chimp" he inspired.

In connection with Zippy's appearance, each store will sponsor a coloring contest. The children who submit the 25 best entries will receive prizes at a party in their honor.

At press time, the schedule of the tour had not been completed. The dates set so far are: April 6 and 8, May Company, Cleveland; April 10, 11, Mandel Bros., Chicago; April 13, Famous-Barr, St. Louis; April 15, 16, 17, May Company, Los Angeles; April 18, 19, 20, Macy's, San Francisco. Other dates are being arranged to cover the three months of May, June and July.

Cosmopolitan Zippy doll ad

In March of 1957, Cosmopolitan Doll & Toy Corporation announced Zippy's USA tour to promote their new 8-inch tall Zippy doll, along with their "Ginger" doll.

Ginger (from Cosmopolitan) and Barbie (from Mattel) dolls had come out at the same time, but each had her own appeal. Barbie was sexier, while Ginger was more "wholesome" and "average" in appearance. Ralph and Kathleen were the couple who performed the shows with Zippy at each of the scheduled Ginger appearances. Zippy's USA tour proved to be very profitable, to the point that Ralph was able to buy a plane and fly to each location. Unfortunately for Cosmopolitan, Barbie's appeal outlasted Ginger's. The Cosmopolitan Zippy doll had human proportions and was not as attractive and cuddly as the Rushton Zippy.

Some other merchandising items, produced as a result of the Howdy Doody Show, were a Zippy hand puppet, coloring books and paint sets. There were many additional merchandising items, most of which we were not aware of at the time. The Kagran Corporation produced and held rights to all merchandising resulting from the Howdy Doody Show and went wild with hundreds of items. Unfortunately for us, our contract did not include royalties from these products, (or from reruns), so they were more or less meaningless to us. Just out of curiosity, I would like to see some of these products. NBC and Kagran benefited royally.

I once read an article about "Buffalo" Bob Smith, who passed away in North Carolina in 1998. He was worth over seventeen million dollars. I guess he profited well from the merchandising. We did receive royalties on the Zippy Doll from Rushton for about ten years.

Early in Zippy's career, Zippy and his dad were having lunch in the very exclusive Roosevelt Hotel Restaurant. The president of Rand McNally was sitting at the next table and asked if he could do a children's book about Zippy. The contract was signed then and there, and Zippy the Chimp, Zippy Goes to School and Zippy's Birthday Party Elf Books were born. These were the only books in the Rand McNally line that used real photographs instead of drawings. Ben Mitchell was the photographer and remained a true friend for many years thereafter.

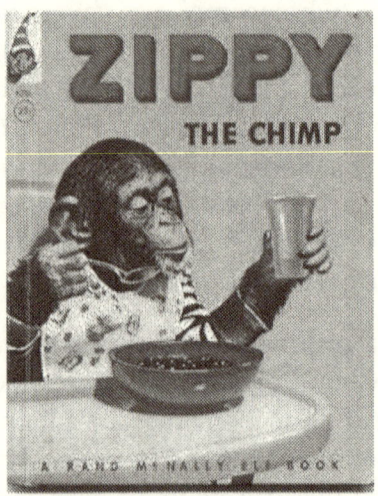

First Zippy the Chimp Elf book
(Rand McNally publisher)

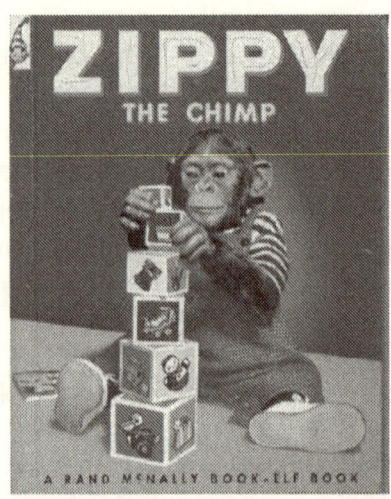

Second Zippy the Chimp book

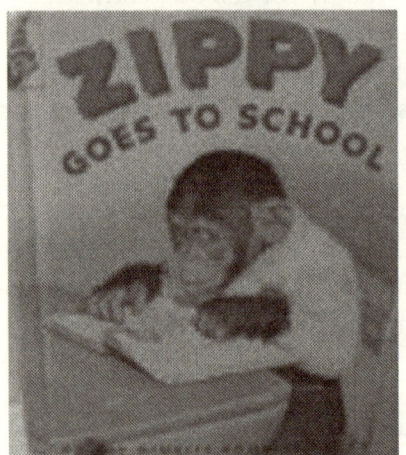

Third Zippy the Chimp book
(Goes to School)

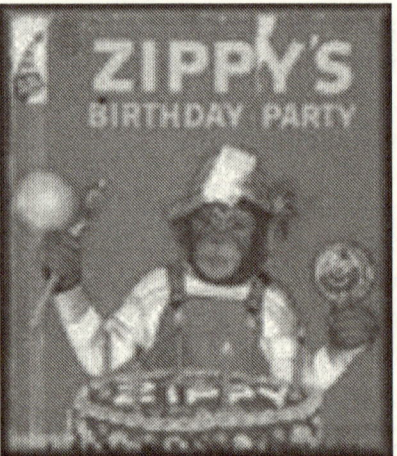

Fourth and last Zippy book
(Birthday Party)

The original <u>Zippy the Chimp</u> book's cover had Zippy sitting in a high chair eating from a bowl. This book is rare, because only a few of this version were ever printed. The second edition portrays Zippy playing with blocks. Later, after the royalty contract was over, Rand McNally printed a very large Zippy book. Who knows what else is out there.

Castle Films produced two 8mm home movies of Zippy: <u>Zippy Goes to the Beach</u>, and <u>Zippy's Birthday Party</u>. The more expensive ones were in color, while the less expensive films are shorter and in black and white. I was still in my teens when these were made, and yes, that's me portraying Zippy's "mother." If you are looking for Zippy memorabilia, there were only two official Zippy films made. Later, Castle Films picked up some other chimp movies and sold them with the Zippy name on them. In filming <u>Zippy Goes to the Beach</u>, Zippy kept getting sand in his shoes, so I spent a lot of time carrying him piggyback.

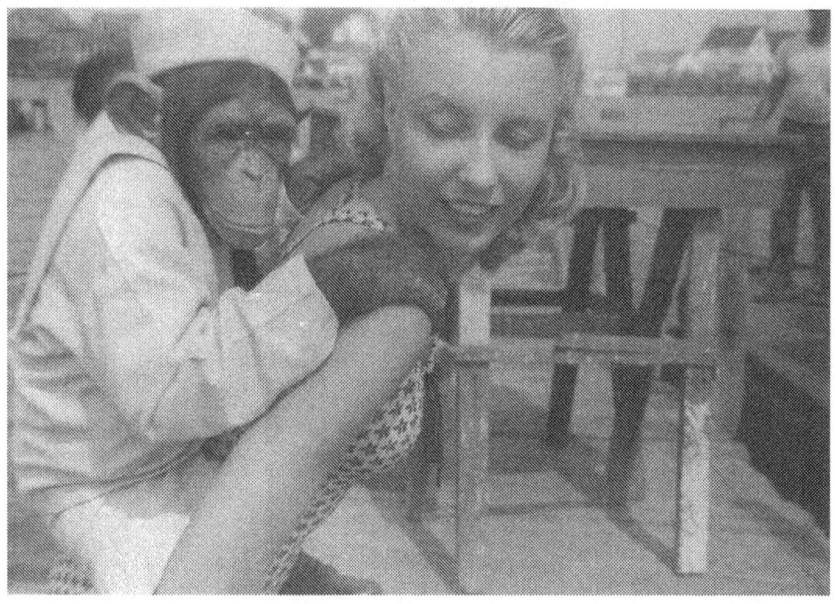

The sand was too hot for Zippy's feet

After we were signed on with CBS Literary Enterprises, Inc., Pines Comics published the Zippy comic books. As in the Elf Books, I am portrayed as Zippy's "mother." The two children in the comics are totally fictitious. I never had children; maybe that's why I loved the chimps so much. The cartoon artwork is great, and the stories are fun. Only two of these comics, (numbers 50 and 51), were completed before our contract with CBS ended. They have become highly collectible.

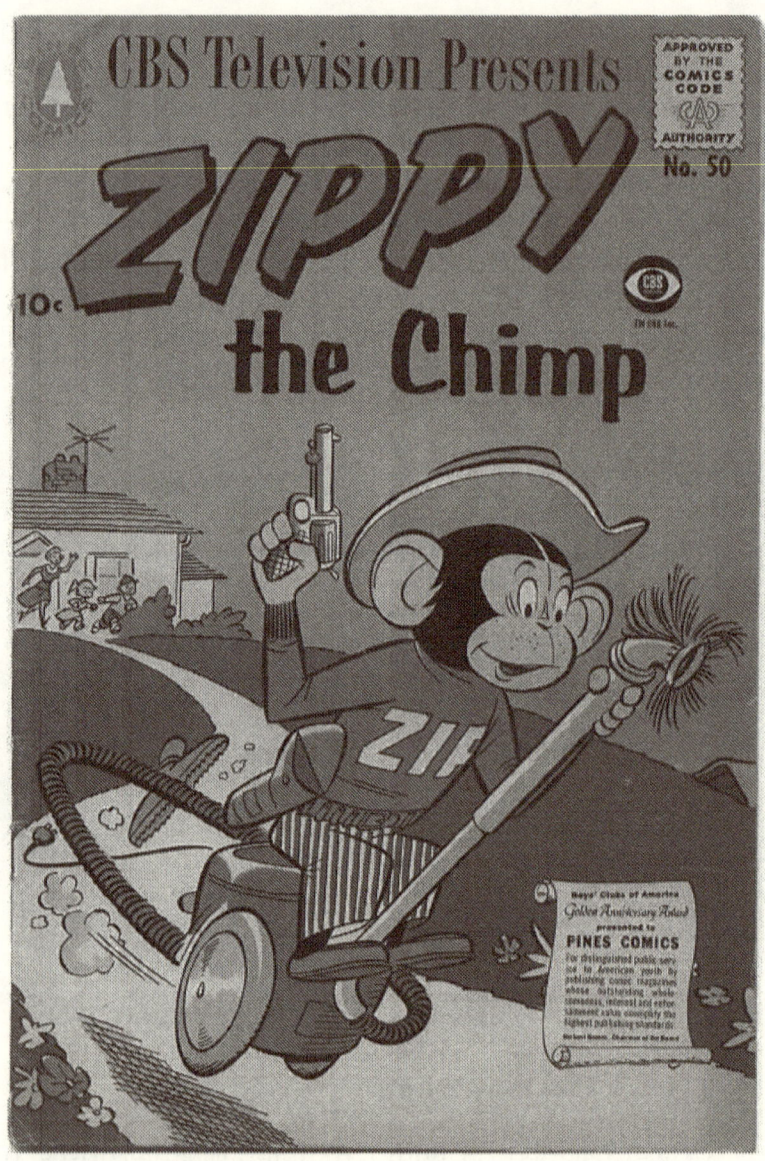

#50 Zippy comic (CBS/Pines Publications)

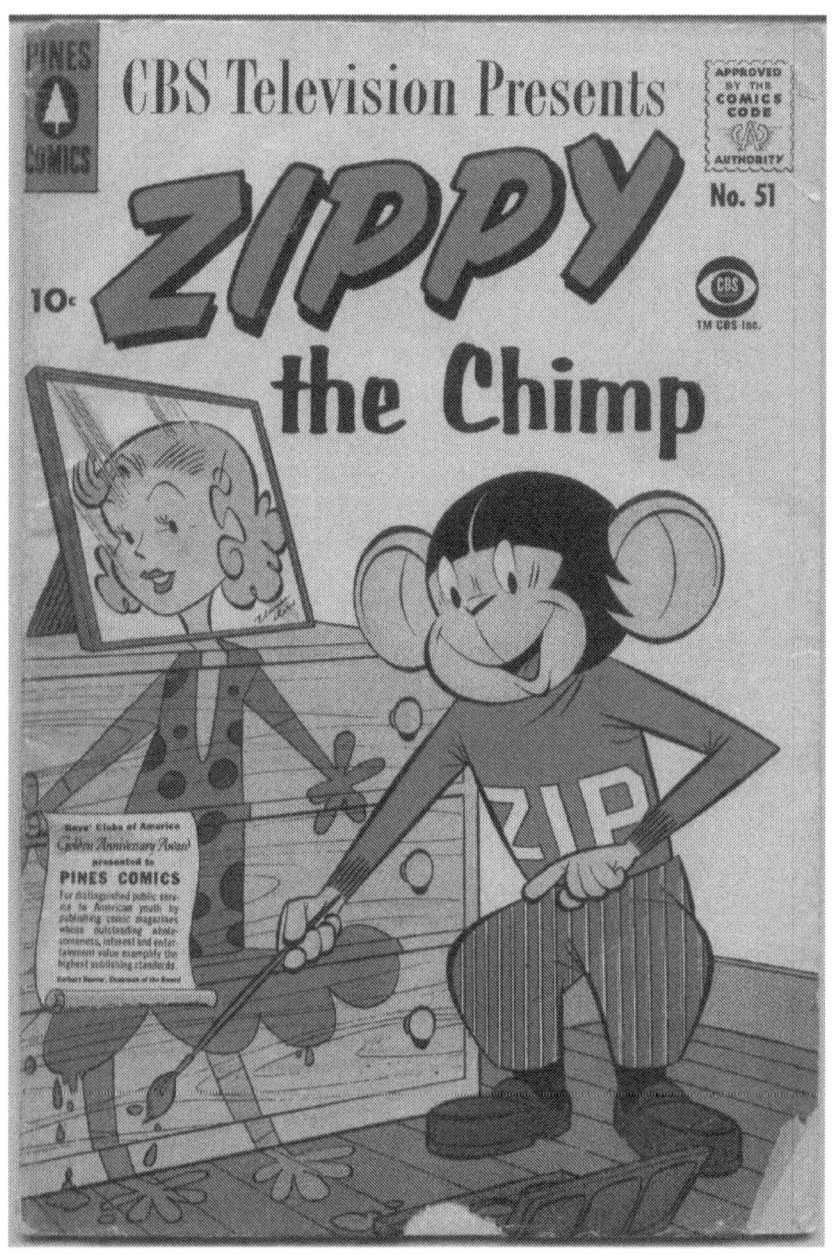

#51 Zippy comic (CBS/Pines Publications)

# To My Zippy Fans...

To reiterate what I mentioned earlier, I am looking forward to a second book about Zippy! The next book will be about experiences other people have had with Zippy. If you met him on the street, had him perform at a party, or he somehow made an impression on you, tell me about it. If you are one of the people who worked with Zippy (an agent, one of the people I trained or someone who shared the stage) or if a Zippy Doll was one of your best friends, tell me about it. Pictures will be most welcome and a signed release allowing us to use your material is a must. Funny, entertaining and uplifting stories are preferred.

Please write and send materials to:

Zippy the TV Chimp

PMB 12283

P.O. Box 2430

Pensacola, Florida 32513

www.zippythechimp.com

www.zippytheTVchimp.com

www.ingramcontent.com/pod-product-compliance
Lightning Source LLC
Chambersburg PA
CBHW022001170526
45157CB00003B/1099